Analyzing the Hazard Evaluation Process

Kenneth L. Dickson
Alan W. Maki
John Cairns, Jr.

Editors

Proceedings of a Workshop
held in Waterville Valley, New Hampshire
August 14–18, 1978

Water Quality Section
American Fisheries Society
Washington, D.C.
1979

Copyright 1979 by the
Water Quality Section,
American Fisheries Society

Library of Congress Catalog Card Number: 79–92254

All orders should be addressed to

American Fisheries Society
5410 Grosvenor Lane
Bethesda, Maryland 20014 USA

CONTENTS

v Dedication

Introduction

A. W. Maki K. L. Dickson John Cairns, Jr.	1	Introduction

Hazard Assessment Approaches

C. M. Lee	7	Determination of the Environmental Acceptability of Detergent Components
P. Lundahl	23	Hazard Assessment in Schemes for New Chemicals in France
J. Gareth Pearson John P. Glennon	30	A Sequential Comprehensive Hazard Assessment Strategy for Estimating Water Quality Criteria
K. Fujiwara	50	Japanese Law on New Chemicals and the Methods to Test the Biodegradability and Bioaccumulation of Chemical Substances
R. Lloyd	58	The Use of the Concentration-Response Relationship in Assessing Acute Fish Toxicity Data
J. R. Duthie et al.	62	Discussion Synopsis—Hazard Assessment Approaches

Hazard Assessment Philosophy and Principles

James W. Akerman David L. Coppage	68	Hazard Assessment Philosophy: A Regulatory Viewpoint
W. Brock Neely	74	An Integrated Approach to Assessing the Potential Impact of Organic Chemicals in the Environment
Alan W. Maki	83	An Analysis of Decision Criteria in Environmental Hazard Evaluation Programs
Eugene E. Kenaga	101	Aquatic Test Organisms and Methods Useful for Assessment of Chronic Toxicity of Chemicals
Donald I. Mount	112	Adequacy of Laboratory Data for Protecting Aquatic Communities
R. A. Kimerle et al.	119	Discussion Synopsis—Hazard Assessment Philosophy and Principles

Water Quality Criteria

Kenneth J. Macek
Sam R. Petrocelli
    122   A Fisheye View of Water Quality Criteria

Jerry L. Hamelink   127   A Proposed Method for Deriving Effluent Limits from Water Quality Criteria

Dana J. Davoli
Eileen R. Choffnes
    132   The Development of Water Quality Criteria—An Environmentalist's Viewpoint

Leonard J. Guarraia
John J. Carroll
Kenneth M. Mackenthun
    138   Data Needs in Developing Water Quality Criteria

H. E. Johnson et al.   143   Discussion Synopsis—Water Quality Criteria

Summary and Conclusions

C. M. Fetterolf et al.   148   Summary and Conclusions

    157   Index

Dedication

Lloyd Lyman Smith, Jr.
1909–1978

This symposium volume is dedicated to the memory of Dr. Lloyd L. Smith, Jr., a friend and colleague of many of the participants. Lloyd was a steadfast member of the American Fisheries Society and a widely respected research investigator in aquatic toxicology. He was one of the pioneers in fish bioassays whose research was always characterized by the highest professional standards and attention to detail. One could place confidence in his results and expect in his papers a full and thorough disclosure of all important details, including identification of the weaknesses as well as the strengths. We had hoped to have Lloyd play a key role in the symposium on which this volume is based. The news of his death was a shock, particularly because it preceded this symposium by only a few weeks. The sadness due to his loss is tempered by the knowledge that his zest for research and problem solving in aquatic toxicology did not diminish with age. One of the first symposium decisions was that this volume would be dedicated to Lloyd as a salute to a respected colleague whose contributions to the field will endure through his publications and his students.

INTRODUCTION

Introduction

A. W. Maki, K. L. Dickson, and John Cairns, Jr.

The chemical industry has experienced rapidly expanding growth over the last 75 years to the point where there are now approximately 2 million recognized chemical compounds and derivatives in existence with almost 250,000 new compounds synthesized each year. While providing numerous societal benefits, modern chemical technology has also brought about a marked increase in the number and variety of chemical pollutants ultimately entering natural surface waters. Recent estimates from the United States Environmental Protection Agency indicate that although the majority of these new compounds will not be commercialized, nearly 1,000 new chemicals annually have the potential for environmental release as a result of marketing, use, and disposal. As our understanding of the structure and functioning of the aquatic environment has grown, it has become obvious that the capacity of our rivers, lakes, and oceans to assimilate chemical wastes and toxic materials discharged into them is not at all infinite and that serious degradation of water quality is the inevitable result of misuse and mismanagement of this invaluable resource.

In today's society, the classification of all chemicals as distinctly harmful or beneficial is not warranted, and we recognize that it is not feasible to describe a strict line of demarcation between harmful and beneficial. It is more relevant to recognize that there exist degrees of harm and safety for all chemicals since even the most innocuous of chemicals can create distinctly harmful environmental effects when present in high concentrations. In contrast, even the most toxic chemical substances can be assimilated by the aquatic environment provided the concentrations are sufficiently low. Therefore, it is evident that all chemical substances can be considered to have potential for toxic effects on surface-water communities, but that the degree of harm or safety of a chemical substance must be related to the amount potentially reaching the aquatic environment.

Along with the future societal benefits from technological developments, we must be able to predict changes in water quality and quantify impacts on aquatic life. We recognize that no chemical can ever be pronounced safe; however, *risk* associated with its use may be judged acceptable. An immediate need exists to develop further expertise and understanding of existing freshwater and marine aquatic communities so that an unacceptable risk to the survival potential of resident aquatic life will not be imposed. There is also a need to predict, assess, and separate unimportant and unavoidable impacts of toxic substances on these communities from those changes and impacts that are significant and degrading. Methods to derive these needed data are currently evolving as procedures for *hazard evaluation*. This is defined as the development of methods for measuring risk to the aquatic environment associated with the use of chemical substances through an objective and probabilistic exercise based on empirical data and scientific judgment. The results of the integrated hazard evaluation procedure can then be used as a uniform approach to provide a means of estimating the limiting concentrations of chemical substances which will produce no observable effects on aquatic life potentially exposed to the chemical.

Background

In response to a recognized need for the development of relevant testing methodology and integrated testing programs for effectively and efficiently assessing the potential hazards to aquatic life associated with the use of chemical substances, an ad hoc group planned and organized a workshop on *Estimating the Hazard of*

Chemical Substances to Aquatic Life, held June 13-17, 1977 at Pellston, Michigan.

After reviewing methods and techniques of aquatic toxicology and the basic components of the hazard assessment process, the Pellston participants recognized that it would be beneficial to hold a subsequent workshop which examined additional hazard assessment approaches from this country as well as approaches developed in Europe and Japan. Therefore, the scope of this second workshop would include a consideration of the philosophy, development, and informational needs of programs for evaluation of hazard to aquatic life associated with the use of chemical substances. In addition, the participants of the Pellston workshop identified two additional topics of concern related to evaluating the hazards of chemical substances to aquatic life for consideration in a subsequent workshop. These topics were:

(1) How well do laboratory-derived toxicity data predict ecosystem effects?

(2) What constitutes an adequate data base for water quality criteria for chemical substances?

It was concluded that a subsequent workshop addressing these concerns would be of significant assistance to professionals in industry, government, and environmental groups responsible for dealing with these problems.

In order to implement this mandate from the participants of the Pellston Workshop, an ad hoc planning committee meeting was held in October 1977 at the American Society for Testing and Materials meetings in Cleveland to plan the 1978 Waterville Valley Workshop. The names of that planning committee were as follows:

Dr. John Cairns, Jr. &
  Kenneth L. Dickson
Virginia Polytechnic Institute
  and State University
Blacksburg, Virginia

Dr. Jerry L. Hamelink
Lilly Research Laboratory
Greenfield, Indiana

Dr. Howard E. Johnson
Michigan State University
East Lansing, Michigan

Dr. Richard A. Kimerle
Monsanto Industrial Chemical Company
St. Louis, Missouri

Dr. Kenneth J. Macek
Bionomics, EG&G, Inc.
Wareham, Massachusetts

Dr. Alan W. Maki
Procter & Gamble Company
Cincinnati, Ohio

Dr. Foster L. Mayer, Jr.
Fish-Pesticide Research Laboratory
Columbia, Missouri

Dr. Charles R. Walker
US Department of the Interior
Washington, D.C.

The committee was responsible for the identification of specific program topics for the workshop sessions and recommended the list of participants:

Dr. Karim Ahmed
Natural Resources Defense Council
New York, New York

Mr. James Akerman
US Environmental Protection Agency
Office of Pesticide Programs
Washington, D.C.

Dr. George Baughman
U.S. Environmental Protection Agency
Athens, Georgia

Dr. Wulf Karl Besch
Institut fur Wasser-und
  Abfallwirtschaft
Karlsruhe, West Germany

Dr. Wesley Birge
University of Kentucky
Lexington, Kentucky

Dr. William E. Bishop
Procter & Gamble Company
Cincinnati, Ohio

Dr. Dean Branson
Dow Chemical Company
Midland, Michigan

Dr. William Brungs
US Environmental Protection Agency
Duluth, Minnesota

Dr. John Cairns
Virginia Polytechnic Institute
  and State University
Blacksburg, Virginia

Ms. Eileen Choffnes
Citizens for a Better Environment
Chicago, Illinois

Mr. David Coppage
US Environmental Protection Agency
Office of Pesticide Programs
Washington, D.C.

Dr. Dana Davoli
Citizens for a Better Environment
Chicago, Illinois

Dr. N. T. deOude
Procter & Gamble Company
Brussels, Belgium

Dr. Kenneth L. Dickson
North Texas State University
Denton, Texas

Dr. Tom Duke
US Environmental Protection Agency
Gulf Breeze, Florida

Mr. James R. Duthie
Procter & Gamble Company
Cincinnati, Ohio

Mr. Carlos M. Fetterolf, Jr.
Great Lakes Fishery Commission
Ann Arbor, Michigan

Mr. Mack Finley
US Fish & Wildlife Service
Columbia, Missouri

Professor Kikuo Fujiwara
The University of Tsukuba
Niihara-Gun, Ibaraki-Ken, Japan

Dr. Leonard Gaurraia
US Environmental Protection Agency
Washington, D.C.

Dr. Jerry L. Hamelink
Lilly Research Laboratory
Greenfield, Indiana

Mr. David Hansen
US Environmental Protection Agency
Gulf Breeze, Florida

Dr. Peter Hodson
Canada Centre for Inland Waters
Burlington, Ontario, Canada

Dr. Paul Hull
Gilbert/Commonwealth
Reading, Pennsylvania

Dr. Curtis Hutchinson
Union Carbide Corporation
Tarrytown, New York

Dr. Joseph F. Jadlocki
FMC Corporation
Princeton, New Jersey

Dr. Howard E. Johnson
Michigan State University
East Lansing, Michigan

Dr. Anne Jones
Colorado State University
Fort Collins, Colorado

Dr. Eugene E. Kenaga
Dow Chemical Company
Midland, Michigan

Dr. Richard A. Kimerle
Monsanto Industrial Chemical Company
St. Louis, Missouri

Dr. C. M. Lee
Unilever Research Laboratory
Merseyside, England

Dr. G. Fred Lee
Colorado State University
Fort Collins, Colorado

Dr. Richard Lloyd
Salmon & Freshwater Fisheries Laboratory
London, England

Dr. Gordon Loewengart
Allied Chemical Company
Morristown, New Jersey

Dr. Pierre Lundahl
Institute National de Recherche
Chimique Appliquée
Paris, France

Dr. Ken Macek
Bionomics, EG&G, Inc.
Wareham, Massachusetts

Dr. Alan W. Maki
Procter & Gamble Company
Cincinnati, Ohio

Dr. Donald I. Mount
US Environmental Protection Agency
Duluth, Minnesota

Dr. Brock Neely
Dow Chemical Company
Midland, Michigan

Mr. Rod Parrish
Bionomics, EG&G, Inc.
Pensacola, Florida

Dr. J. Gareth Pearson
US Army Medical Bioengineering Laboratory
Frederick, Maryland

Dr. Sam Petrocelli
Bionomics, EG&G, Inc.
Wareham, Massachusetts

Dr. Richard Purdy
US Environmental Protection Agency
Washington, D.C.

Dr. James Reisa
US Environmental Protection Agency
Washington, D.C.

Dr. Phillip W. Schneider, Jr.
E.I. DuPont
Newark, Delaware

Dr. Gil Veith
US Environmental Protection Agency
Duluth, Minnesota

Dr. Charles R. Walker
US Department of the Interior
Washington, D.C.

Scope and Objectives

The following workshop objectives were presented to the participants in order to direct their efforts:

(1) To critically review and analyze the state of the art in assessing the hazards of chemical substances to aquatic life.
(2) To critically review and analyze the state of the art in developing water quality criteria and subsequent recommendations for safe levels.
(3) To evaluate the efficacy of laboratory-derived data to predict natural ecosystem effects.

In an introductory statement of challenge to the workshop participants, Carlos Fetterolf, Jr., of the Great Lakes Fishery Commission, underlined the need to utilize these objectives to focus specific discussions during the workshop sessions. Participants were also challenged with the responsibility of representing their individual professions and scientific disciplines in their contributions to the workshop rather than the particular organizational or institutional viewpoints associated with their jobs. Participants were encouraged to identify specific aspects of the workshop topics which they endorse as professionals and believe are relevant to the development of hazard evaluation procedures for the protection of aquatic life.

To address the above specific objectives, six individual workshop sessions were identified within the following outline:

Session 1: Introduction

Perspective and relation to the 1977 Pellston Workshop. A discussion of the utility and impact of the workshop effort. Presentation of goals and objectives and challenge to the participants.

Session 2: Harzard Assessment Approaches

Presentations from several facets of industry and governmental agencies detailing currently employed hazard assessment schemes, discussion of utility, and actual case study examples.

Session 3: Hazard Assessment Philosophy

(A) Identification of a specific "basic data set" required for decisions and risk analysis in environmental hazard evaluation programs. Is the development of such a basic series of tests desirable, and if so, what test data (physical/chemical and biological) should be included?
(B) An in-depth assessment of specific *decision criteria* used in several existing hazard evaluation programs to arrive at decisions regarding further testing, to use or to discontinue use of a particular substance. What role do physical/chemical data play in the decision making? How many test species are sufficient, from which trophic levels?
(C) Extrapolation: how well can laboratory test data be extrapolated to real-world environmental exposures? Do laboratory-generated safety factors and Maximum Allowable Toxicant Concentrations (MATC values) provide for protection of aquatic communities? How well do results of modeling studies predict chemical partitioning of test materials?
(D) Case studies: A presentation of actual case studies of several real chemicals, ideally to include examples of materials currently in use, undergoing testing, and discontinued for reasons of unacceptable risk. Can consistent trends in testing programs be demonstrated? What similarities and differences in approaches are evident?

Session 4: Water Quality Criteria

What data base is needed to establish acceptable, scientifically valid, and ecologically relevant water quality criteria for chemical substances? What is the relative impact of chemical and biological data on the establishment of criteria?

Session 5: Committee Workday

On Thursday, the participants worked in committee to prepare synopses of assigned sessions. Under the leadership of the chairman of each session, the responsibility of each committee was to prepare a short synopsis of each session including important conclusions or recommendations which had the consensus of the committee.

Session 6: Workshop Summary

Session committees presented their conclusions and recommendations. An overall workshop summary was prepared by the workshop summary committee.

Workshop Conduct

The workshop was held at the White Mountains Conference Center located in Waterville Valley, New Hampshire. The site was selected as providing a reasonable balance between accessibility and remoteness to encourage informal discussions and evening sessions among participants. On Sunday, August 13, 1978, workshop participants arrived at the lodging headquarters, the Landmarc Inn, and were each given a loose-leaf notebook containing final drafts of each of the discussion initiation papers and a copy of the published proceedings of the 1977 Pellston conference.

Each workshop session was initiated with one or more well defined discussion initiation paper solicited from a specific author or group of authors named by the ad hoc committee. These papers were presented during plenary sessions of the workshop followed by a discussion period open to all participants under the leadership of a previously appointed chairman. It was the responsibility of these chairmen and identified committees to transcribe each plenary discussion session to consensus conclusions on Thursday of the week. Each committee then presented its findings and conclusions to a plenary session of all workshop participants on Friday.

A transcript was made of each discussion session by court recorders and these transcripts were made available to the session chairmen by the end of the day to be used in summarizing the individual sessions. Transcripts were made available to the chairmen immediately so that the discussion of plenary sessions could be reviewed and incorporated into the individual committee summations. The session chairmen and their committees met on Wednesday afternoon and evening and throughout the day and evening on Thursday in order to produce a consensus summary of each session for presentation to the plenary session on Friday morning. During the Friday plenary session, initial drafts of consensus summaries were presented and the final drafts were completed by the session chairmen through subsequent mailings to all members of respective synthesis committees. However, no substantial changes were permitted after the workshop was finished. The transcription of the plenary session responses were reviewed by the editors to produce this introductory section. With this one exception, and the discussion initiation papers prepared before the workshop convened, all other material represents a distillation of the discussion sessions which followed the presentation of the discussion initiation papers. Every attempt has been made to ensure that editorial review has been carried out in the spirit of the workshop. Since the consensus views were the primary objective of the workshop, it was deemed inappropriate to attempt anything more than editorial changes following the termination of the workshop.

Acknowledgments

The financial support of the following industries via a grant-in-aid to Virginia Polytechnic Institute and State University made this workshop possible and is gratefully appreciated:

Allied Chemical Company
Dow Chemical Company
E.I. Du Pont de Nemours and Company
Eli Lilly Company

Monsanto Chemcial Company
Procter & Gamble Company
Unilever Company

We wish to thank Mr. Rick Blauvelt and the staff of the Waterville Valley Resort Association for their hospitality and innumerable contributions toward the success of the workshop. We are also indepted to Earnest W. Nolin & Associates, General Stenographic Reporters, Manchester, New Hampshire, for their efficient handling of the daily transcripts. Verbatim transcripts of the individual discussion sessions were made available to session chairmen by the evening of the same day, thus facilitating the task of the individual synthesis committees. The editors also want to acknowledge with sincere appreciation the efforts of Ms. Darla Donald who helped with the many organizational and editorial aspects of the workshop; Ms. Eva Dickson, who helped with the editorial aspects of the workshop; Ms. Mary Wind and Ms. Terra Clem who typed and collated these proceedings.

HAZARD ASSESSMENT APPROACHES

Determination of the Environmental Acceptability of Detergent Components

C. M. LEE

Bioconsequences Section, Unilever Research Laboratory
Port Sunlight, Wirral, Merseyside, England

Abstract

A hazard evaluation scheme for the components of domestic detergent products is described which is based on predictive testing limited by time and resources. One of the three sequential test programmes, which progress through a series of stages of increasing complexity, is chosen depending on the concentration of the chemical expected in the environment.

The scheme recognizes that laboratory studies, no matter how detailed, cannot prove the complete absence of risk. The objective is, therefore, to indicate the degree of risk to the environment associated with the particular use of a chemical by identifying the safety margin between exposure and effect concentrations.

The case study example of the potential detergent builder trisodium carboxymethyloxysuccinate is used to demonstrate how the exposure of the material, together with the results and implications of biodegradability, treatability, and toxicity tests, determine its progress through the evaluation scheme.

Domestic detergent products enjoy widespread use and are disposed of as aqueous solutions through domestic wastewater systems. Their components are, therefore, likely to be dispersed throughout the aquatic environment, and this presents particular difficulties for the determination of their environmental acceptability. The same components may enter waters of widely differing chemistry and biological community structure through a variety of sewage treatment processes. Neither the manufacturer nor the consumer can control the disposal and subsequent treatment, if any, which domestic detergent products may receive, so that it is essential that all components be compatible with existing treatment systems. In this they differ from chemicals discharged in factory effluents where, if necessary, specific treatment methods can be devised.

Hazard evaluation schemes are based on predictive testing limited by time and resources so that adequate safety margins are needed to predict acceptability. Laboratory studies, no matter how detailed, cannot prove the absence of risk, and the objective of any evaluation scheme is to determine the scale of the risk attached to the use of a particular chemical as an aid to judging its acceptability. The paper which follows presents a scheme for examining detergent components which is still under development.

Principles of the Scheme

The assessment of the risk attached to the use of a chemical must include careful consideration of its likely concentrations at various points along its disposal route and in the open environment, its stability and overall distribution in the environment, and its chemical and biological properties. Whilst a rigid scheme for the assessment of this risk cannot be applicable in all cases, it is possible to develop general guidelines and indicate the criteria upon which such an assessment may be based, whilst retaining sufficient flexibility to allow the proper exercise of scientific judgment.

In many cases, the risk of significant effects in the aquatic environment will be sufficiently small for a decision on acceptability to be reached after a limited amount of testing using, for the most part, established methodology. The scheme presented here has been designed to deal with these relatively straightforward cases and indicate the cases where a more detailed examination is called for. For this reason, if the results of such an examination suggest that the safety margin is too small for acceptability to be assumed, the chemical need not necessarily be rejected since a more detailed examination may allow acceptability to be established.

The scheme is based on the principle that the risk to the environment and hence the amount of testing required is related to, but not exclusively dependent on, the concentrations of the material in the environment. It should therefore be clear that a chemical judged acceptable for one application or level of usage may be judged unacceptable in the other circumstances.

Factors Considered

Level of Usage

The concentrations of a chemical in the environment will be influenced by the proposed level of usage. An estimation of the concentration which is likely to occur in raw domestic sewage in the country or countries of interest, therefore, forms the basis of the scheme. This estimate must incorporate a safety margin and since it is important that the safety margin is not unrealistically large, two calculations are made:

(a) on the basis that *all* products of the type in question (e.g., all toilet soaps, all dishwashing liquids, all fabric washing powders, etc.) contain the given level of compound;
(b) on the basis of the actual amount of chemical likely to be discharged as a direct result of the proposed application. This figure is multiplied by ten to provide a safety margin. The lower of these two concentrations is taken as the calculated level in sewage (CL).

The data needed to make these calculations and to calculate the worst possible case level in the open environment include:

(1) percentage of the component in the product;
(2) consumption of detergents in the area, country, or countries of interest;
(3) water use per capita in the area, country, or countries of interest;
(4) percentage of domestic sewage subject to treatment;
(5) river flows (where relevant);
(6) average dilution of effluent in rivers.

The calculation of likely environmental levels may be refined by a knowledge of biodegradation and removal either during sewage treatment or subsequently in the river water. With new compounds these latter data are not usually available at an early stage in examination and their acquisition may form an important part of the experimental programme.

Chemistry of the Compound

Consideration of the structure and properties of a chemical may suggest specific interactions with other components in the environment. In this respect, comparison with analogous chemical structures that are already released to the environment and whose behaviour is known may be useful. The following points are among those which are taken into account.

Legislation

Does the chemical belong to a class subject to legislative control? For example, anionic and nonionic surfactants must pass the Organization for Economic Cooperation and Development (OECD) confirmatory test of biodegradability if they are to be used in European Economic Community (EEC) countries.

Chemical Purity

It is important to identify the nature and amounts of the major impurities present in commercial samples of the chemical and to consider their environmental acceptability separately. For some materials which are complex mixtures (e.g., most surfactants), this will not be possible. In these cases, it may be necessary to consider the relative rates of biodegradation of the various components and to ensure that aquatic toxicity studies are done on a mix of components which is representative of that likely to occur in the environment.

Stability

Account must be taken of processes such as photolysis, hydrolysis, thermal degradation, etc., which may result in the disappearance of the chemical or the formation of new chemical species in the environment. It will, therefore, be necessary to consider the likely stability of the product ingredient during processing, storage, and use and also in domestic sewage and the receiving environment.

Chelation Properties

For those materials which may chelate heavy metals, it may be necessary to consider their possible impact on the mobilization of heavy metals and the biodegradability of the heavy-metal chelates.

Solubility/Partition Coefficient

Where a material has only limited solubility in water but is soluble in lipids, the possibility of bioaccumulation must be considered.

Nutrient Content

The possible contribution of the compound to the nutrient load of surface waters must be taken into account.

Biodegradability

If a compound undergoes rapid biodegradation, the concentrations in the environment and hence the risk to aquatic life will be significantly reduced. Evidence of biodegradability, therefore, simplifies the task of establishing environmental acceptability, but biodegradability should not be regarded as essential in all cases. By classifying the tests on the basis of the information which they provide, two main classes can be recognized.

(a) *Biodegradability-Potential* tests provide an indication of the inherent susceptibility of a chemical to metabolism by microorganisms. Tests in this class may be further subdivided, according to the conditions they provide for acclimatization and biodegradation to occur, into tests of ready biodegradability or of recalcitrance.

(b) *Simulation* tests provide an indication of the rate of biodegradation under the same environmentally relevant conditions. Tests of this type may be subdivided according to the environment they are designed to simulate, e.g., aerobic or anaerobic biological treatment, river, estuary, sea, soil, etc.

The test conditions will determine to which class a test belongs, but the analytical technique used will also influence whether evidence of bioelimination or of primary or ultimate biodegradation is obtained.

In the evaluation scheme which is described, the biodegradability tests are organized to begin with a simple, stringent test of ready biodegradability, where a positive result may be regarded as unequivocal evidence of biodegradability, but a negative result only indicates that more favorable conditions may need to be provided for biodegradation to occur. More favorable conditions and, if necessary, a longer test period can be provided in a test of recalcitrance to determine if the compound is inherently biodegradable. A negative result in this type of test still need not lead to the conclusion that the compound is non-biodegradable but may indicate that biodegradation in the environment will be slow or unreliable. If adsorption seems likely, it may be useful at this stage to determine the removal of the compound during sewage treatment. In the event of a positive result, the evidence that the chemical will not persist indefinitely in the environment may be all that is required in some cases. However, in those cases where biodegradation is important to the avoidance of adverse environmental effects, it may be necessary to proceed to an appropriate simulation test. This will provide information on the rate of biodegradation under conditions which are relevant to the cases. It will also allow the refinement of the calculation of the environmental concentration which is essential to the consideration of environmental acceptability. The relevance of biodegradability testing and tests to determining the environmental acceptability of organic compounds is described more fully by Gilbert and Watson (1977).

Treatability

Since detergent components are discharged to the environment in domestic wastewater and sewage, they must be compatible with the existing patterns of wastewater treatment and disposal. Unfortunately, the type of treatment received varies from none (direct discharge to river or sea), through septic tanks and percolation fields serving individual houses or small communities, to the large modern sewage treatment plant operating the activated sludge process and anaerobic digestion of sludges. In most cases an examination of the behaviour of the chemical in a simulated activated sludge plant and anaerobic digester will be sufficient, though for a major detergent component a range of processes may need to be considered.

These studies help to refine calculation of likely environmental levels by indicating the proportions of the compound likely to be biodegraded, the amount retained by the sewage solids (which may be dumped at sea or spread on agricultural land), and the levels present in the treated effluent which is generally discharged to river or sea. The distribution of a chemical between seattleable solids and the liquor of raw sewage can be determined in a laboratory test designed to simulate the primary settlement of sewage. A laboratory-scale, activated sludge plant is used to determine the distribution of the chemical between seattleable solids, suspended solids, and liquor so that the amounts going to each step of sewage treatment are known. It is particularly important to know the amount of a chemical that goes to the sensitive anaerobic digestion stage in this type of treatment. The concentration of a chemical which will inhibit the anaerobic digestion of sewage solids may be determined by examining the effect on gas production of batch digestors. Other forms of treatment such as trickling filters, septic tanks, contact-stabilization tanks, etc., and processes such as sludge dewatering and sludge disposal can be examined as the case demands.

Toxicity to Aquatic Life

Sewage treatment will rarely be completely effective in removing even a readily biodegradable material. It must, therefore, be assumed that all chemicals used in detergent products will occur in rivers at some finite level and that their impact on aquatic life must be considered. The crucial issue in judging the acceptability of a chemical in the aquatic environment is the safety margin between the exposure concentration and the effect concentration. The degree of risk associated with this judgment will depend on the confidence limits attached to the observed safety margin as much as the size of the margin.

In this scheme it is regarded as more useful to determine at which concentration of a chemical a given response is produced than to determine the length of time for the response to be produced at a given concentration. It is clearly impossible to examine the toxicity of a chemical to all the aquatic species likely to be exposed to it under all the conditions which might possibly occur in the environment, so representative species have to be selected. A considerable number and variety of test methods have been developed throughout the world using species representative of all levels in the aquatic ecosystem from bacteria to fish. In the scheme described here, test organisms are selected as representatives of the major trophic groups within the aquatic ecosystem. Particular species are chosen which are convenient to culture and handle in toxicity tests and for which there are retrospective toxicity data and whose biology are well known. The option is also kept open to take account of the physicochemical characteristics of the test compound by selecting particularly vulnerable representatives of the aquatic ecosystem; e.g., when the chemical is likely to be strongly adsorbed on solids benthic species might be chosen. As far as possible, species relatively sensitive to toxicants are chosen, which confers additional confidence on decision criteria developed from the test results.

Tests are organized to progress from the simpler, short-term acute tests where death or immobilization are the commonly accepted criteria of response, to the more complex, longer-term chronic tests where sublethal effects are measured. In the latter category, the EIFAC (1975) recognizes six categories of tests, including determination of effects on the biochemistry, histology, physiology, growth rate, life cycle, and behaviour of the organism. It is acknowledged, however, that the implications of effects on reproduction and the life-cycle stages are more easily recognized. Consequently, in the scheme for assessing the effects of detergent components where it may be desirable to use chronic tests, these are confined for the present to determining effects on reproduction and life-cycle.

Clearly, the acute toxicity tests only provide a crude estimate of environmental hazard and, therefore, a large safety margin is required if a judgment on acceptability is based only on this type of information. Sprague (1971) has, for example, suggested that sublethal effects on the organism will be unlikely at concentrations ≤0.01 of the LC50. Information on sublethal effects will reduce this uncertainty and allow the use of a smaller safety margin in establishing acceptability.

Experience has not shown bioaccumulation to be a major problem with detergent components, which may be due in part to the emphasis placed on using water-soluble and degradable materials. The potential for accumulation should first be indicated when the chemistry of the material is considered, e.g., likely high fat solubility and low water solubility. A material which has high stability under hydroytic, light, heat, and microbial conditions should always be considered further with respect to accumulation since its persistence may make it more readily available to aquatic organisms in the receiving water. Measurement of the water solubility and octanol/water partition coefficient, with action standards <0.5 mg/liter^{-1} and >1,000, respectively (AIBS 1977), may be useful indicators of the need to proceed with experimental work.

Operation of the Scheme

The suggested scheme is organized into a number of stages of which the first is a paper exercise to determine the scope of the problems presented by the chemical under review. The three or four subsequent stages embody methods of increasing complexity and cost, and while the stage boundaries are not rigid, they provide an opportunity to assess progress and to estimate likely costs and timing of future work. Progress through these stages is controlled by a series of planning diagrams which are designed to ensure an adequate examination of a chemical by a logical sequence of tests. Each test, or series of tests, is followed by a question to which only a definite yes or no response can be given (test methods used are briefly annotated in the Appendix). Of course, it is not possible in a scheme of this type to anticipate every possible outcome of a test. In some cases, the results obtained may not allow a straightforward interpretation or may raise questions which will not be answered by the scheme. In these cases such evidence should not be ignored and sufficient flexibility must be allowed for such unexpected problems to be investigated and resolved. It is important to remember that where the results obtained fail to establish acceptability, the compound need not necessarily be rejected but it may mean that a more detailed examination is required. The case studies of trisodium carboxymethyloxysuccinate (CMOS), a potential replacement for phosphates in fabric washing powders, will be used to demonstrate the operation of the scheme.

Preliminary Examination

The examination of the environmental acceptability of a potential new component of detergent products begins with a paper exercise (Stage I) in which the literature is reviewed, the chemistry of the compound is considered, and the predicted levels of the various chemical species likely to occur in sewage are calculated (Figure 1). Depending on the calculated level of the chemical in raw sewage, one of the three test programmes is selected (Stage II and beyond) as shown in Figures 2–5.

If the builder CMOS were to replace phosphate in fabric washing powders, levels in the product of 10–25% would be used and it must, therefore, be regarded as a potentially major ingredient. It is the trisodium salt of a relatively simple mono-ether tricarboxylic acid, and sodium carbonate (2.0%), sodium sulphate (1.0%), disodium glycolate, disodium fumarate, and disodium maleate (all <0.5%) are present as impurities. These impurities are all known from the literature to be biodegradable and of low toxicity so that further examination is not considered necessary. The Na$_3$CMOS is stable under normal wash conditions and conditions likely to prevail in the environment, and it exhibits no interaction with sodium hypochlorite bleach. Although CMOS is expected to be found in the environment as the calcium salt, it is regarded as only a moderate chelating agent and is not expected to transport or mobilize heavy-metal ions in the environment to a significant extent.

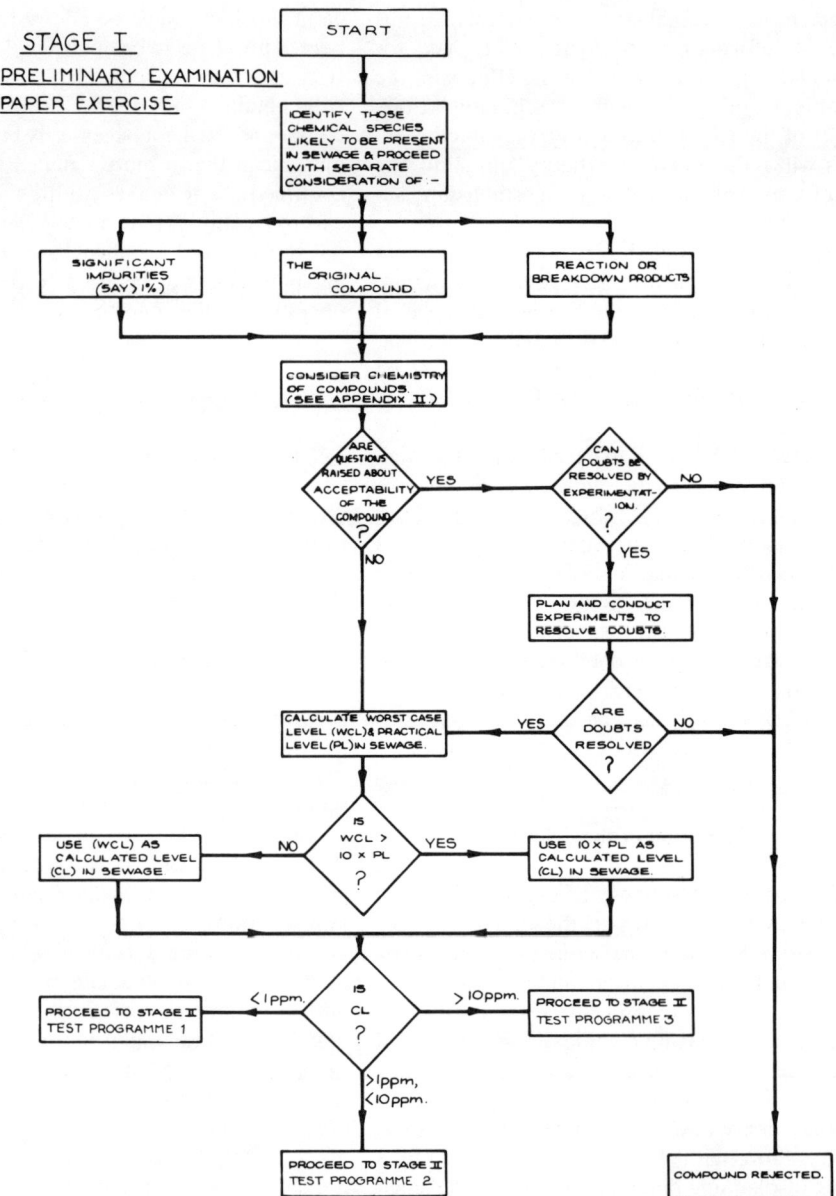

FIGURE 1.—*Stage I analysis for determination of test material concentrations in wastewater.*

Taking account of the tonnage of heavy duty detergents used, an average daily per capita water consumption of 400 liters (domestic use only) in the USA, and assuming that the total heavy duty detergent market included Na_3CMOS at 25% in the product, a worst-possible-case level of 20 mg/liter Na_3CMOS might be expected in raw sewage. Using the same baseline data but assuming a more realistic 10% inclusion of Na_3CMOS in the product for only 15% of the heavy duty detergent market, 1.2 mg/liter Na_3CMOS might be expected in raw sewage. Applying a factor of ten to this latter figure provides a safety margin and gives the calculated level in sewage (CL) as 12 mg/liter, which is used for the purpose of subsequent examination. The third test programme (Figures 4 and 5) would be used in this case. Before discussing the progress of Na_3CMOS through this programme, however, the characteristics of the simpler programmes are described.

for acute toxicity results are based on those of Sprague (1971) and are written in a form which acknowledges the safety margin built into the calculated level (CL), including that attributable to biodegradation where it is demonstrated, but ignoring the potential of subsequent dilution and further degradation of the effluent. In this way, additional confidence is gained for the observed safety margin between likely exposure and effect levels. This approach is also used in Stage III. The use of the question, "any adverse effects?" implies a value judgment of the results of chronic toxicity tests. Whilst it is acknowledged that at present this will tend to mean any statistically significant departure from control values, the increasing sophistication of both test and analysis methods may render this simplistic approach unrealistic. The lack of ability to relate the results of many laboratory chronic tests to field circumstances is currently a major gap in our knowledge.

Test Programme 1:
Calculated Level in Sewage: <1 mg/liter

Environmental concentrations of the class of compounds, which may be minor adjuncts or minor impurities present at a level of less than 1% in fabric washing powders, will be relatively low. These compounds are, therefore, assumed to be acceptable if they can be shown to be biodegradable and to have relatively low acute toxicity to aquatic life (Stage II, Figure 2). For biodegradable materials, only when the concentration of the chemical which is acutely toxic to aquatic life is within one order of magnitude of the estimated concentration in domestic sewage will it be necessary to proceed to Stage III. In this event a limited amount of work on the chronic toxicity of the compound to aquatic life may be necessary.

A non-biodegradable compound may also be judged acceptable if removal in an activated sludge plant is better than 80%, provided that the calculated levels of the compound do not inhibit anaerobic digestion and can be shown not to affect seed germination or aquatic life in simple tests.

In Stage II of the test programme, a compound is considered biodegradable if there is a ≥90% loss in a die-away test or a Semi-Continuous Activated Sludge test (SCAS). The decision criteria

Test Programme 2:
Calculated Level in Sewage: >1–<10 mg/liter

The potential environmental concentrations of chemicals falling into this class, i.e., materials present in fabric washing powders at levels of approximately 1%–10%, will be such that biodegradation will play an important role in reducing them. Evidence of ultimate biodegradability is, therefore, desirable in all cases.

In Stage II, therefore, the biodegradability and acute toxicity of the compound are determined. If the compound is biodegradable and not acutely toxic to aquatic life at one tenth of the calculated level in sewage, the examination proceeds to Stage III (Figure 3). At this stage, the estimates of the likely levels of the compound in treated effluent and adsorbed on sewage solids are refined on the basis of the results of an activated sludge test, and the estimates of effluent level (EL) are then used to determine the significance of the results of subsequent experiments.

Assessment of the possible impact on aquatic life proceeds with the determination of the chronic toxicity to algae, *Daphnia,* and chironomids if acute effects were observed at concentrations within two orders of magnitude of the effluent level. If these shorter tests suggest that the safety margin is large (e.g., no effects at 10× effluent level), then the longer

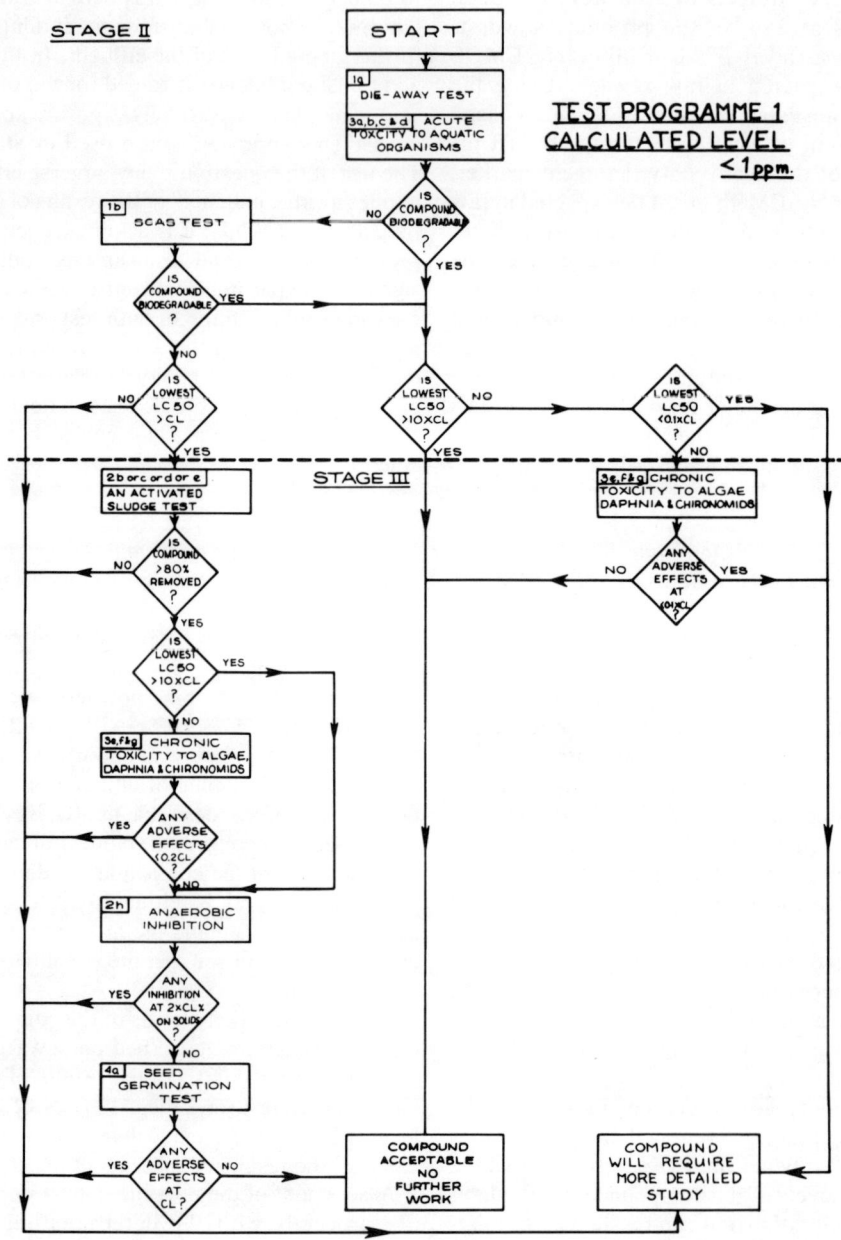

FIGURE 2.—*Stage II testing for initial predictions of fate and effects of the test compound.*

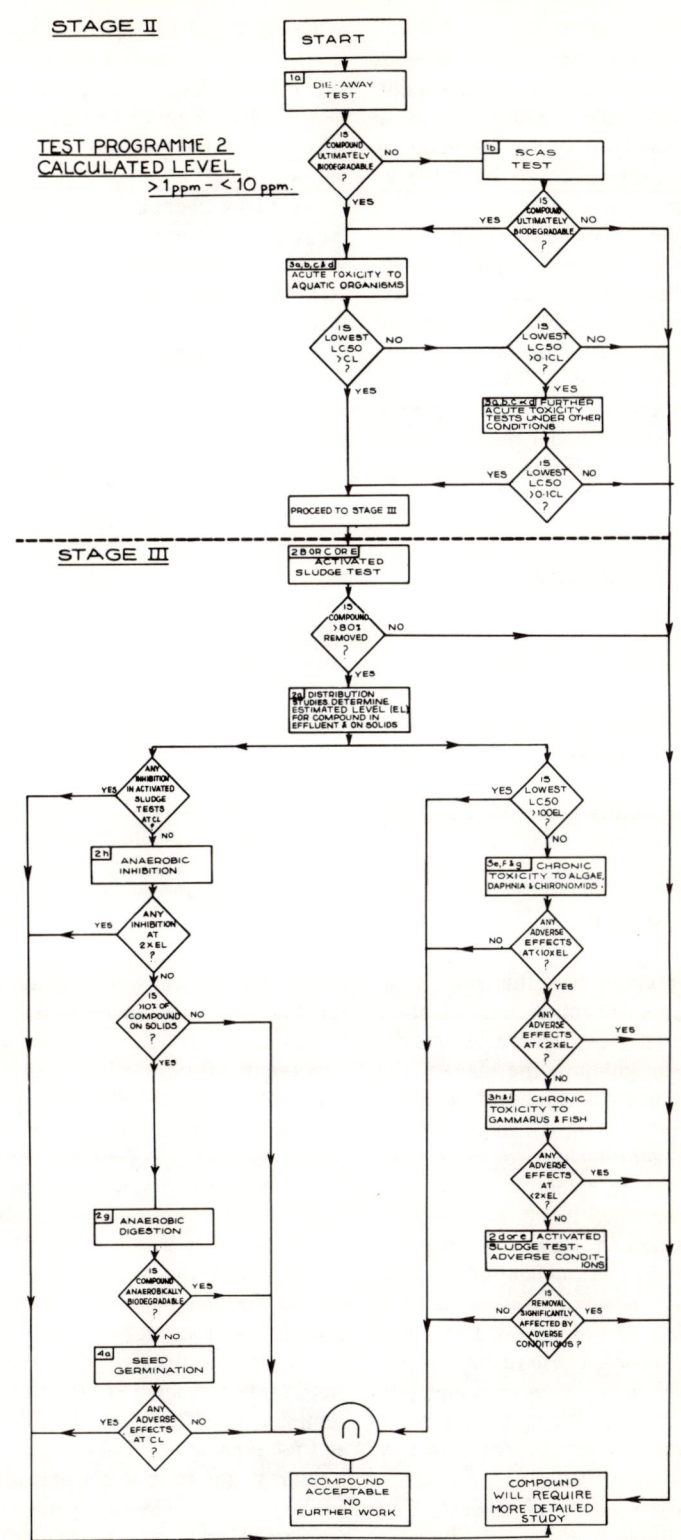

FIGURE 3.—*Stage II testing for ultimate biodegradability and chronic toxicity.*

tests on fish and freshwater shrimps need not be undertaken. If, however, the safety margin is smaller than this, not only should these additional tests be made but also it will be important to check that the treatability of the compound is not disproportionately affected by adverse operating conditions.

At the same time, it will be necessary to examine the effect the material might have on the treatment and disposal of sewage sludges. Anaerobic inhibition is, therefore, examined and, if a significant proportion of the compound is associated with the sewage solids, its impact on sludge dewatering is assessed. Unless a compound is shown to degrade under the conditions of anaerobic digestion, it will also be necessary to assess its effect on seed germination.

Test Programme 3:
Calculated Level in Sewage: >10 mg/liter

Compounds in this class are likely to be present in fabric washing powders at levels >10% and are potential major environmental contaminants which will therefore receive close scrutiny (Figures 4 and 5). The results of the Stage I examination clearly place Na_3CMOS in this class.

Stage II

As in the early programmes, this stage consists of simple but stringent biodegradability tests and acute toxicity tests where the results will give the first insight into the size of the safety margin which might be expected, and also the scale of the problems which might be encountered in demonstrating environmental acceptability. In a simple river die-away test, 20–100 mg/liter Na_3CMOS exhibited complete primary biodegradation in 4 days at temperatures in the range 4–24°C. Complete mineralization of Na_3CMOS was demonstrated using a radio labelled sample in a die-away test when 91% $^{14}CO_2$ was recovered. These results clearly show that Na_3CMOS is rapidly and completely biodegradable at concentrations of the same order as the calculated levels expected in raw sewage.

In simple acute toxicity tests, the lowest LC50 was 300 mg/liter to *Daphnia magna* in soft water (25 mg/liter as $CaCO_3$) which is more than an order of magnitude greater than the calculated level in raw sewage. Because of the apparent size of the safety margin, confirmed by toxicity test results for other species, it was not necessary to enter the final loop of the Stage II examination.

For many chemicals the results obtained for Na_3CMOS under the Stage II examination would have been more than adequate to demonstrate environmental acceptability without further work. However, since Na_3CMOS is potentially such a major ingredient, a full examination was judged necessary.

Stage III

In order to refine estimates of the likely exposure levels, an activated sludge test is done to determine the level of removal during sewage treatment. This is followed by distribution studies to determine the level of the compound on sewage sludges and in the effluent. In the case of Na_3CMOS, a laboratory-scale, activated sludge plant run on domestic sewage was used to show that removal was >95% at both 15°C and 5°C within 7 days, when the inlet level of Na_3CMOS was 20 mg/liter. Since the compound is so water soluble and so little would be expected to adsorb on solids, distribution studies were thought to be necessary. This decision was justified by the subsequent demonstration that Na_3CMOS is completely anaerobically biodegradable. Using an anaerobic sewage digestor operated under conditions found in a municipal sewage treatment plant, up to 20 mg/liter Na_3CMOS was >95% removed after 6 days acclimatization. This result suggests that even in the unlikely event of all the calculated level of Na_3CMOS in raw sewage going with the solids to the digestor, no treatment problems would be encountered. It was also on the basis of these results that it was decided that it was not necessary to examine the effects of Na_3CMOS on the dewatering of sewage sludges.

Taking account of the proposed use of Na_3CMOS at a level of 10% in a product representing 15% of the heavy duty detergent market, and its expected >95% removal during sewage treatment, the level in sewage effluent would be ~0.06 mg/liter. This does not take account of possible dilutions of the effluent which, in the USA, might be 1:10 with river water.

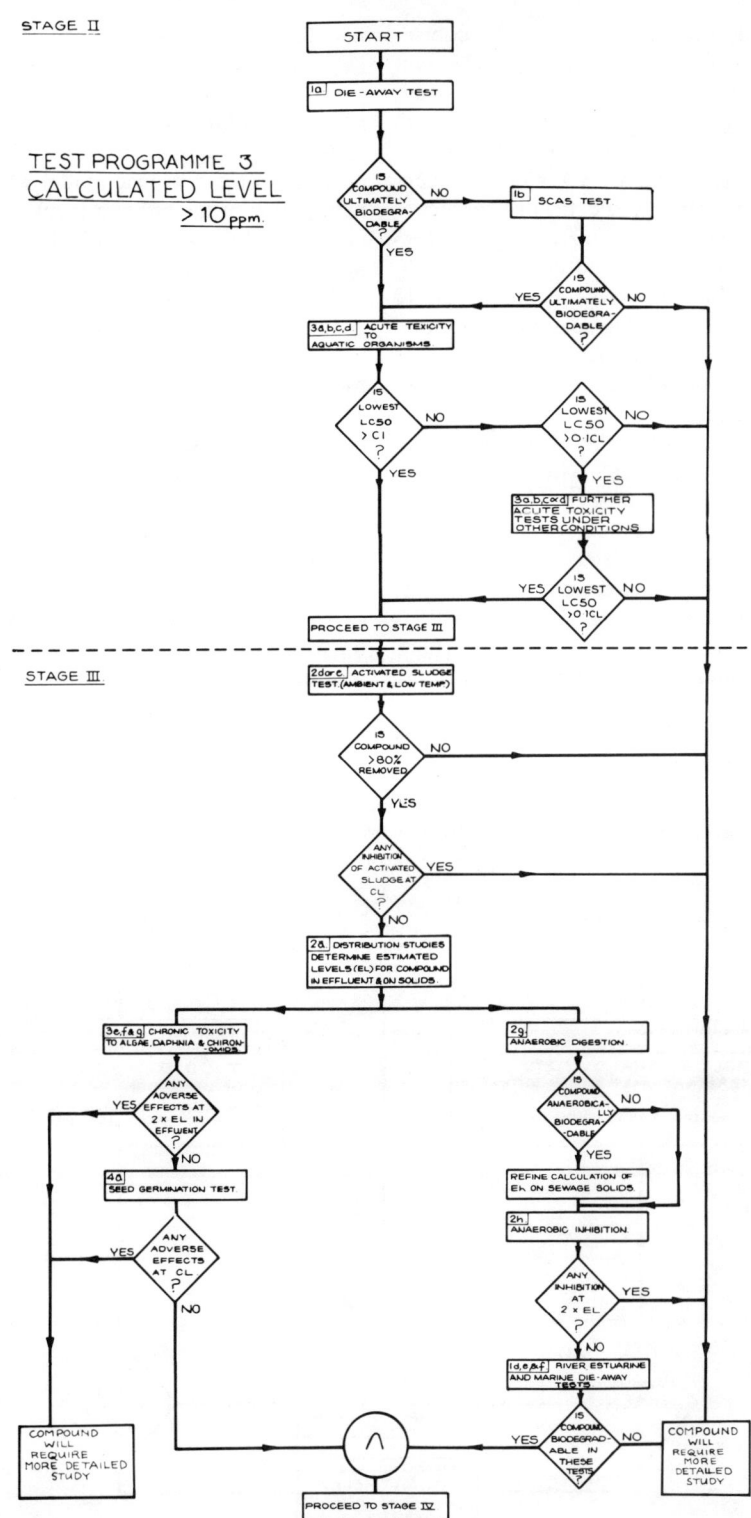

FIGURE 4.—*Advanced Stage III testing for major environmental contaminants.*

It has already been noted that there is considerable variation in the type of sewage and extent of treatment given to domestic sewage. Indeed, domestic sewage may receive no treatment at all, particularly in communities close to estuaries or the sea, and since significant concentrations of Na_3CMOS could be discharged to these coastal marine and estuarine waters, die-away tests under marine and estuarine conditions are desirable. In a river die-away test 10 mg/liter Na_3CMOS was degraded in 9 days, and in a similar die-away test using estuarine water (1–4% salt content) 10 mg/liter Na_3CMOS degraded in 29 days.

Using the refined effluent level as a guide, the shorter chronic toxicity tests with invertebrates are done to determine if an adequate safety margin is maintained when sublethal effects are examined. Chronic toxicity tests with algae, *Daphnia*, and chironomids did not show any significant effects attributable to Na_3CMOS at levels $\leq 2 \times$ effluent level (0.12 mg/liter). There was no stimulation of algae at up to 100 mg/liter, no effects on survival or production of juvenile *Daphnia* at ≤ 50 mg/liter, and no effects on the survival, growth or reproduction of chironomids at <100 mg/liter Na_3CMOS (Lee et al., in review). The safety margin is, therefore, considerable (2–3 orders of magnitude) when the possible dilution of the effluent is taken into account and, though smaller, would be adequate even under direct discharge conditions.

Since waters containing Na_3CMOS might be used for the irrigation of crop plants, its effects on the germination of pea, barley, radish, and tomato were examined at 13.9 and 139 mg/liter. There was a reduction in germination success of radish (25%) and tomato (40%) at 139 mg/liter Na_3CMOS. The germination of the seeds of the other species were unaffected at 139 mg/liter and there were no effects on any plant species at 13.9 mg/liter Na_3CMOS. Since this latter concentration is 3 orders of magnitude greater than the expected sewage effluent level, it is unlikely that the irrigation of crop plants with waters containing Na_3CMOS will cause problems.

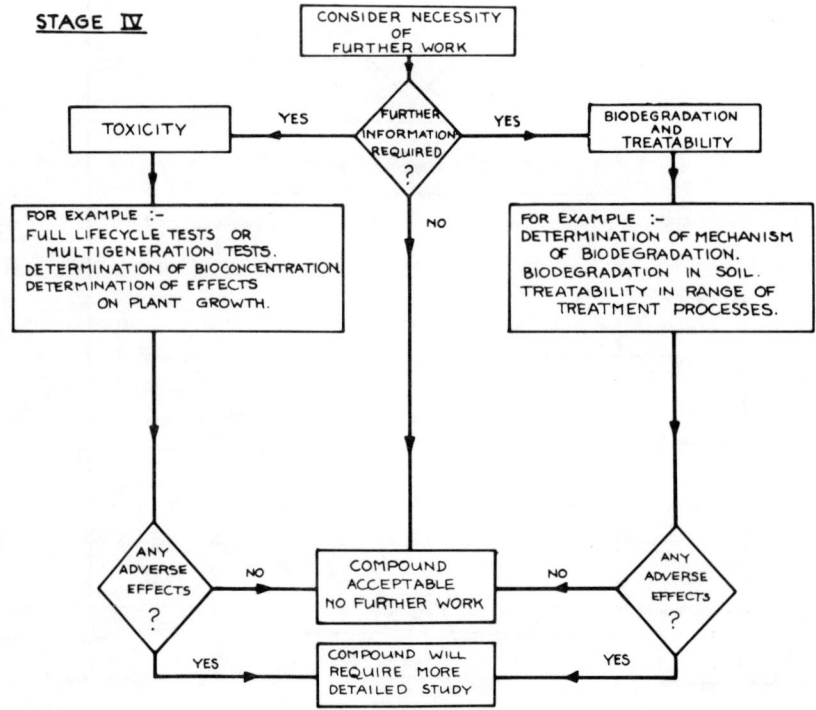

FIGURE 5.—*Stage IV testing needs for long-term or chronic effects.*

Stage IV

The prospects of establishing the environmental acceptability, even of a major component, will usually be clear by the end of Stage III. At the beginning of Stage IV there is an opportunity to consider the necessity of any further work by determining the potential significance of any special problems. Rigorous progression through a series of tests cannot be specified at this stage and experiments will tend to be research projects designed specifically for the case under consideration. With a major component such as Na$_3$CMOS which reaches this stage of the examination, it is essential to show that during degradation no refractory or toxic intermediates are produced. In this case its metabolic pathway was established to determine the nature of the intermediates:

$$\begin{array}{c} CH_2COO^- \\ | \\ CHCOO^- \\ | \\ O \\ | \\ CH_2COO^- \\ \downarrow \\ CHCOO^- \\ | \\ ^-OOCCH \\ \text{Fumarate} \\ + \\ OHCH_2COO^- \\ \text{Glycolate} \\ \downarrow \\ CH_2COO^- \\ | \\ HOCHCOO^- \\ \text{Maleate} \\ \downarrow \\ CH_2COO^- \\ | \\ O{=}CCOO^- \\ \text{Oxalacetate} \\ \downarrow \\ CH_3COCOO^- \\ \text{Pyruvate} \end{array}$$

A new and previously undiscovered enzyme was found to catalyze the degradation, and the reaction produces only intermediates that are rapidly and readily metabolized by both microbial and animal cells and that are normally present metabolites (Peterson and Llaneza 1974). No intermediates are, therefore, likely to persist in the environment.

The high polarity and hydrophilic character of Na$_3$CMOS precludes absorption in fatty tissues, and, therefore, bioaccumulation is unlikely. Multigeneration chronic toxicity tests with *Gammarus pulex* and the convict cichlid (*Cichlasoma nigrofasciatum*) showed no effect on growth or survival up to 100 mg/liter and, in the case of the convict cichlid, no effects on the production and survival of the F$_1$ generation.

A plant growth test using the same representative crop plants as in the germination test indicated only slight stimulation of growth, which may have been due to the marginally increased availability of mineral nutrients by the chelating power of the Na$_3$CMOS. These minor effects were, however, at 139 mg/liter Na$_3$CMOS, which is 3 orders of magnitude greater than the exposure levels.

The effects of Na$_3$CMOS on the operation of trickling filters have not yet been determined, but it has been shown that while it is only poorly degraded during the short residence time in a septic tank, it is well degraded in the drainage field even after the latter has been ponded (Klein and Jenkins 1972).

This completes Stage IV of the examination, and in view of the results which show that a safety margin of several orders of magnitude between the exposure level and the effect level of Na$_3$CMOS is maintained under a wide variety of conditions, the chemical must be regarded as acceptable for the proposed use.

Acknowledgments

The author wishes to pay tribute to Dr. P. A. Gilbert, whose ideas have played a significant part in the development of this scheme, and to other colleagues in Unilever Research for their helpful advice and discussion.

References

AIBS (AMERICAN INSTITUTE OF BIOLOGICAL SCIENCES). 1977. Criteria and rationale for decision making in aquatic hazard evaluation. Pages

214–273, *in* J. Cairns, Jr., K. L. Dickson, and A. W. Maki, eds. Estimating the hazards of chemical substances to aquatic life. ASTM STP 657. Am. Soc. Test. Mater., Philadelphia.

EIFAC (EUROPEAN INLAND FISHERIES ADVISORY COMMISSION). 1975. Report on fish toxicity testing procedures TC24. 25 pp.

GILBERT, P. A., AND G. K. WATSON. 1977. Biodegradability and its relevance to environmental acceptability. Tenside Detergents 4:171–177.

KLEIN, S. A., AND D. JENKINS. 1972. The fate of carboxymethyloxysuccinate in septic tank and oxidation pond systems. Sanit. Eng. Res. Lab., Col. Eng. School Publ. Health, Univ. Calif., Berkeley, SERL Rep. 72. 10 pp.

LEE, C. M., J. F. FULLARD, AND E. HUNTINGTON. In review. The sublethal effects of trisodium carboxymethyloxysuccinate on the midge *Chironomus riparius* (Miegen): the development of a chronic toxicity test. ASTM (Am. Soc. Test. Mater.) J. Test. Evaluat.

PETERSON, D. R., AND J. LLANEZA. 1974. Identification of a carbon-oxygen lyase cleaving the ether linkage in carboxymethyloxysuccinic acid. Arch. Biochem. Biophys. 162:135.

SPRAGUE, J. B. 1971. Measurement of pollutant toxicity to fish, III: sublethal effects and "safe" concentrations. Water Res. 5:245–246.

Appendix: Test Methods

The various test methods which will provide information relevant to the operation of the scheme are briefly described below. The list is not intended to be immutable and it is anticipated that certain tests will be replaced as new methods are developed. If a test does not appear here, this should not be taken as an indication that it is regarded as unnecessary or irrelevant in all cases. The tests selected are those thought to have the widest relevance to the relatively straightforward cases covered by the scheme; in certain situations other information may be required.

Biodegradability Tests

Die-Away Test

Disappearance of dissolved organic carbon (DOC) or evolution of CO_2, from a mineral salts medium in which the test compound is the sole source of carbon, is measured to provide an indication of ready, ultimate biodegradation.

SCAS Test

Semi-continuous activated sludge (SCAS) equipment is fed with sewage and test compound for a period of up to 3 months. DOC is used to follow biodegradation of the test compound. The test is useful for the examination of materials which require prolonged acclimatization before biodegradation occurs.

OECD Screening Test

This is a die-away test in which specific analytical technique is used to follow the course of biodegradation. Information on ready, primary biodegradation is provided.

River, Estuarine, and Marine Die-Away Tests

The biodegradation of the compound in samples of water taken from the environment is followed. These tests may provide an indication of the rate and extent of biodegradation of the compound under environmentally relevant conditions.

Anaerobic Biodegradability Test

No routine test is at present available for determining anaerobic biodegradability, though ^{14}C-labeling techniques may be used and some information is also provided by anaerobic inhibition tests (see "Anaerobic Digestion," below).

Soil Biodegradation Test

Biodegradability by soil microorganisms may be followed either in a closed bottle test (O_2 uptake) using a soil innoculum or in a soil percolation test using ^{14}C-labeled test compound.

Mechanism of Biodegradation

Information on the intermediates produced during the biodegradation of a chemical may be obtained from a detailed study of its mechanism of biodegradation by a pure culture of microorganisms.

Treatability Studies

Distribution Studies

The distribution of the chemical between settleable solids and the liquor of raw sewage is determined in a laboratory test designed to simulate the primary settlement of sewage. A ^{14}C-labeled sample of the chemical will be required or a specific analytical technique.

OECD Confirmatory Test

New evidence of >80% removal of anionic surfactants (by methylene blue analysis) or nonionic surfactants (by Wickbold analysis) is or shortly will be required by law in EEC and other European countries. The test can be used to predict the behaviour of other chemicals under activated sludge conditions if a suitable analytical technique is available.

Coupled Units Test

This is a modification of the OECD confirmatory test in which DOC measurements are used to follow the removal of the compound. Since the DOC method is applicable to all organic compounds, the test should have wide applicability and is claimed to measure ultimate biodegradability. However, the possible loss of original compound or its metabolites by adsorption cannot always be excluded.

Porous Pot Test

This is a further variation of the laboratory activated sludge test which is simpler to operate than the confirmatory test. Since a feed of detergent free sewage or domestic sewage can be used, the tests can also be run at reduced temperatures to simulate winter conditions. A specific analytical technique or ^{14}C-labeled sample will generally be required.

LASP Study

A laboratory-scale, activated sludge plant (LASP) is designed to simulate as closely as possible the actual activated sludge treatment process. Removal of the compound under a range of typical operating conditions may be followed over a period of several months by means of a specific analytical technique. Information may also be provided on the effect of the compound on the treatment process (inhibition) by comparing the rates of DOC and NH_3 removal with those of control units.

Trickling Filters

Pilot scale equipment may be used to determine the removal of a chemical during its effects on this mode of biological treatment of sewage.

Anaerobic Digestion

^{14}C-labeling techniques or measurement of the volume of CO_2/CH_4 evolved may be used to determine the biodegradation of chemicals under conditions simulating the anaerobic digestion of sewage.

Anaerobic Inhibition

The concentration of chemical which will inhibit the anaerobic digestion of sewage solids may be determined by examining the effect on the gas production of batch digesters.

Septic Tanks

Pilot scale equipment may be used to determine the behaviour of the chemical in a septic tank and associated percolation field.

Dewatering of Sewage Sludges

It may be necessary to determine the effects of certain chemicals on the settling and dewatering characteristics of primary, activated, and anaerobic digester sludges.

Aquatic Toxicity Tests

Acute Toxicity—Algae

The concentration of chemical which is inhibitory to the growth of a unicellular, green algae *(Chorella vulgaris)* is determined. This organism provides an example of a primary producer in the aquatic ecosystem.

LC50—*Daphnia* (48-Hour); *Gammarus* (72-Hour); Rainbow Trout (96-Hour)

The acute toxicity of the compound to the herbivore *Daphnia magna*, the detritus feeder *Gammarus pulex*, and the carnivore *Salmo gairdneri* is determined in these tests.

Chronic Toxicity—Algae

The effect of the compound on the rate of growth of cultures of *Chlorella vulgaris* is studied over a 2-week period.

Chronic Toxicity—Chironomid; *Daphnia; Gammarus*

The effect of sublethal concentrations of the compound on the survival, growth and reproduction of these representative aquatic invertebrates is studied over a complete life cycle.

Chronic Toxicity—Fish

A full life-cycle test may also be run on fish using, for example, the convict cichlid or the fathead minnow *(Pimephales promelas)*.

Bioconcentration—Invertebrates

The accumulation of the chemical by an aquatic invertebrate relevant to the particular case being examined, and its subsequent loss in a clean environment, could be studied in such a test.

Bioconcentration—Fish

The uptake, metabolism, and excretion of the chemical by the goldfish *(Carassius auratus)* is determined.

Toxicity to Plants

Seed Germination Test

Solutions of the test compound are used to irrigate seeds of various crop plants sown in compost (e.g., lettuce, tomato, radish, pea, and barley). The effect of the chemical on the emergence of seedlings is then noted.

Plant Growth Test

Various crop plants (e.g., lettuce, tomato, radish, pea, cabbage, sugar beet, and potato) irrigated by direct application of solutions of the test compound to the soil or by overhead spraying are examined for symptoms for phytotoxicity over a period of 3 months.

Hazard Assessment in Schemes for New Chemicals in France

P. LUNDAHL

Institut National de Recherche Chimique Appliquée
BP 1, 91710 Vert le Petit, France

Abstract

There is no standard scheme for new chemicals hazard assessment in France. Moreover, there is no generally recognized philosophy. An act was promulgated on July 12, 1977 to protect man and his environment against the dangers which may arise from chemicals but it does not describe the information that will be required. The decrees of application which will describe the test program of the qualification process have not yet been published. The author describes his philosophy and the procedure he suggests to use for chemical substances that may be discharged into freshwater ecosystems. He shows that the results of the preliminary assays, and other information, should be used to determine the amount and kind of complementary information that is needed. A standard scheme could be extremely useful, but a rigid pattern of bioassays would be a waste of time and money. The decision to discard or commercialize a product can often be arrived at without doing a very large number of tests. Toxicity is probably the most important factor but many other possible detrimental impacts on the ecosystem should be taken into account in the decision process.

Environmental awareness has developed strongly during the recent decades as a response to the degradation of the environment. After a period during which the zero-growth policy seemed the only alternative to the steady degradation of the biosphere, a will to develop a new kind of growth which is more compatible with natural equilibriums has appeared. That new growth demands assessment of the environmental impacts of all kinds of projects (new dams, new manufactures, new products, new uses of existing products, new technologies, etc.).

The National Environmental Policy Act (NEPA), the Toxic Substances Control Act (TSCA) in the USA, the Act on the protection of the nature, and the Act on the control of chemicals in France show well that new trend. As the TSCA and the Act on the control of chemicals are still very recent, as are the acts in the few other countries that already have them, there is still a strong need for the development of a methodology in that field.

The Act on the control of chemicals in France was promulgated on July 12, 1977. Its aim is to protect man and his environment against the dangers which may arise from chemicals, and especially from new chemicals. The act does not describe the amount and kind of testing that will be required in the qualification program. These will probably be described in the decrees of application which have not yet been published.

There is no standard scheme or even a standard philosophy for the qualification of new products in France, and this paper presents only the personal views of its author.

In this paper, only the potential impacts of new chemicals on aquatic ecosystems will be considered, but the philosophy of the assessment would be the same for terrestrial ecosystems and for a global environmental assessment.

Unfortunately, an aquatic ecosystem is a very complicated system. An action on only one of its components can disturb the whole system because it can be transmitted and even amplified by the links which connect the components. This is one of the reasons which make it so difficult to assess the potential impact of a chemical on an aquatic ecosystem.

Discussion

Usually, when the assessment of the hazards of chemicals to the aquatic environment is talked about in France, the discussion is immediately centered on toxicity bioassays or biodegradability testing. This is a very dangerous limitation of the scope of the assessment that can easily lead to wrong decisions. Many other possible impacts should be considered, such as inhibition of oxygen transfer through the water surface, modification of sedimentation, foul taste, eutrophication, etc.

The first step in a successful assessment is to determine its purpose. In this paper, it is assumed that the assessment is done to obtain information that is used together with other information, to determine if a product can be commercialized and used without restrictions, or if it can be used under restrictions, or if it should not be used at all. The decision process itself is not discussed, but its relation to the information collection scheme is included. Besides the process of information gathering, some of the assays and measurement methods are examined. The assessment should be done in a way that minimizes the risk of getting "false positives" (acceptance of harmful products), and that at the same time minimizes the risk of getting "false negatives" (discarding of acceptable products). It should also be as inexpensive and fast as possible. Obviously, it is impossible to implement a no-risk policy in that field as in many others. As it is impossible to avoid all "false positives," the decision to produce and commercialize a product should never be considered as irreversible. However, this cannot be seen as an insurance against "false positives" because the social and financial costs of a "false positive" may be extremely high. It is very difficult to explain to the public that a product has been accepted because the risks of harmful consequences of its use are low enough. The average citizen, who, without hesitation, takes the train or plane where he is exposed to accidents with a known probability, often wants absolutely safe products. This is another reason to use information gathering schemes and methods that minimize strongly the probability of getting false positives.

To summarize, it can be said that the purpose of the information gathering process is the determination of the acceptability of a new product by the aquatic environment. The aquatic environment is perceived very differently by its different users; for example, the swimmer first notices the transparency (or the turbidity) and odors, the fisherman first notices the abundance of fish, or if he is a dry-fly fisherman, of aquatic insects. The naturalist notices a rare species. The average citizen notices the taste of tap water, and if he eats fish, he possibly is afraid of the possible bioaccumulation of pollutants in the flesh. The road engineer looks at the biological corrosion of the concrete foundations of the piers of the bridge, etc. All these people look at different components of the aquatic environment and all have legitimate rights to get protection for the components they are interested in.

Moreover, even if some modifications of the aquatic ecosystem are obviously detrimental, such as the destruction of a fish population, others are much more difficult to classify. For example, the partial destruction of aquatic macrophytes that are used for spawning by sunfish in a brown trout river will usually improve the trout fishery (competition is reduced), but it will have a negative impact on a coarse-fish fishery. So, what can be said of traces of herbicides? This is a problem because a new chemical has to be assessed from a very general standpoint, without taking into account the difference between the ecosystems it will contaminate. In this context, a modification that can be either positive or negative has to be regarded as a risk, that is to say, as possibly negative. The consequence of these considerations is that in the hazard assessment process, the ecosystem should be considered thoroughly; the biocenosis, the biota, their relations, the relations between the ecosystem and man, or between the ecosystem and other ecosystems should be considered (Duthie 1977).

To assess the hazards of a chemical in the aquatic environment, the quantity that will reach the aquatic environment and its distribution (in time and space) should be estimated. This quantity is determined by the quantities that are produced, and by the kind of uses. It also depends on what is done with the product after use. Obviously, the case of a transformater fluid which is used only in closed systems and has a very long life-time is very different from the case of a household detergent. It is also obvious that the use of the same set of toxicological assays and of the same rigid scheme to draw conclusions would lead to unrealistic or false decisions. A difficulty is that all the possible uses of a new chemical are not known when the commercialization begins. This is the reason why the French law on the control of chemicals states that if a new danger may arise owing to a change in the quantities marketed, in the manufacturing process, to the ways in which the substance is distributed or used, or to its dispersal in the environment, any producer or importer must make a new declaration to the appropriate administrative authority. The producer or importer may then be subjected to an obligation to provide additional information on the effects of the substance on man and the environment. Usually, the premarketing studies which have been made by the manufacturer, and the knowledge of the kind of commercialization which he intends to make, allow an assessment of the first foreseeable utilizations by types and quantities.

These data are then used to forecast the penetration of the chemical in aquatic ecosystems. The total amount which will enter the aquatic ecosystems is broken down by time and type of location. The following checklist can be used:

Total annual production
  1st year
  2nd year
  Foreseeable production growth after 2nd year
Utilizations
  In closed systems only
  Professional only
    Less than 200 users in the country
    More than 200 users in the country
  Public
    Concentrated in some regions
    Homogenous

Penetration into aquatic ecosystems
  Utilization does not lead to usual discharge into sewage water
  Utilization leads to discharge into sewage water
  The chemical enters the environment by other ways, and
    May finally enter aquatic ecosystems
    Will not finally enter any aquatic ecosystem

Surface waters
  They will receive the chemical in quantities that are distributed in time and space as urban and common industrial sewage
  Different distribution of locations
  Chronic but irregular entry into surface waters
  Seasonal or occasional entry into surface waters
  Accidental entry into surface waters

Ground waters
  Foreseeable chronic contamination
  Accidental contamination only

The questions are answered by "yes" or "no." Then the quantities entering the aquatic environment and their distribution in time and place are determined using the knowledge of such properties as the ability of the chemical to be extracted from solid wastes deposits by leaching, its treatability in sewage waters treatment plants (which depends on its biodegradability, but also on its adsorbability on the sludge), data on the percentage of sewage waters that are treated, the stability of the chemical, its volatility from water, its ability to react with other components of the sewage water, etc. The knowledge of climatological data may also be necessary when examining, for example, a new rubber additive which could be used in tires and washed from the roads into rivers. If the chemical reacts with other substances or is degraded before entering the aquatic environment, the characteristics and quantities of its reaction or degradation products should be studied (except if they already known as harmless).

Usually a precise determination of the quantities which go into the aquatic environment is impossible and the hazard assessment process then deals with orders of magnitude only.

The next step in the hazard assessment is the identification of the most critical accidental and especially, of the most critical non-accidental situations. For example, for a detergent, they could be small rivers downstream of large cities without purification plants. The foreseeable level of concentration is the only factor determining what is called here a "critical situation." The consequences of this level, for example, the possible toxic effects, are taken into account in a later step. The concentrations that will be found in these critical non-accidental situations are estimated by a mere division of the quantity entering the aquatic environment by the available amount of water.

A similar calculation is usually nearly impossible for accidental situations. Except if the probability is especially high or if the consequences of an accident would be especially severe because of the chemical's nature, accidental situations should not be dealt with by decisions limiting production or use but by security measures during the manufacture and the transportation of the product. These accidental situations will not be discussed anymore in this paper. If it can be shown that the product never reaches the aquatic environment, it is considered as acceptable as far as the aquatic environment is considered and no further studies in this field are made.

If the estimate of the concentration in the most critical non-accidental situations shows without further testing of the product that it will have unacceptable effects, it must be discarded and further studies would be a loss of time and money.

At each step of the hazard assessment process, products that show unacceptable effects are immediately discarded. The problem is that in some cases it is far from obvious to determine what is unacceptable. Usually, a chemical is considered as unacceptable if the toxicity studies show that there will be significant modifications of the ecological equilibrium of the aquatic ecosystem, even, of course, if these modifications are the result of a slow evolution, without mortality. It would probably be better to consider as unacceptable any product for which it is impossible to show that it has no significant effect on the ecological equilibrium. But, unfortunately, such a demonstration is usually impossible because there always could be an unforeseeable consequence. Aquatic ecosystems are so complex and diverse that even field tests do not allow a perfectly thorough view of the impact of a pollutant.

To initiate a testing scheme, a series of toxicity bioassays are made. Toxicity thresholds to different species are often very different. The lowest is the most significant, but it is impossible to test all the species of all the freshwater ecosystems (or even of one freshwater ecosystem). To optimize the ratio "relevant information per amount of work" very different types of organisms should be chosen. At our Institut, *Brachydanio rerio* (a small fish), *Daphnia magna* (a crustacean), *Chlamydomonas variabilis* (an alga), and sometimes heterotropic bacteria are used. The 24-hour LC50 on *Brachydanio rerio* in static conditions (AFNOR 1977a), and the 96-hour LC50 on *Brachydanio rerio* in a continuous flow system, the concentration which immobilizes 50% of the *Daphnia* in 24 hours (AFNOR 1974), the concentration which immobilizes 50% of the cells of *Chlyamydomonas variabilis* (Lundahl 1974), the concentration which doubles the generation time of a culture of *Chlorella vulgaris* or *Chlamydomonas variabilis*, and the concentration which inhibits the uptake of traces of radiolabeled glucose by the bacteria of river waters are determined. The ultimate biodegradability using, if possible, the Association Française de Normalization method which is based on the measure of dissolved organic carbon (AFNOR 1977b), is evaluated. If the biodegradability is equal to or higher than the biodegradability of glucose measured in the same assay, if all the toxicity thresholds are well above (more than 10 times) the critical concentrations in the aquatic environment, if the toxicity to mammals is low, and if known properties do not suggest any harmful effect (bioaccumulation risks which are estimated from partition coefficients and affinity for proteins, odor, taste, oxygen transfer, etc., are considered), the product is qualified for a field test and will probably be acceptable (Figure 1).

If the structure of the product raises any doubt about the biodegradability of all the molecule's skeleton, or if the measured biodegradability is significantly lower than the biodegradability of the glucose, a screening for possible hard biodegradation products should be made. Concurrently the toxicity of the biodegradation intermediates should be studied.

The screening for hard intermediates can usually be done by using a ^{14}C-Labeled sample of the chemical. The toxicity of the biodegradation intermediates is determined without purification of the metabolites, according to Lundahl (1974).

In all cases of low biodegradability or slow biodegradation of hard biodegradation products, of bad partition coefficients or affinity for proteins, the risks of bioaccumulation or of accumulation in the sediments are studied. Radiolabeled samples are used in a completely artificial food chain, with *Chlamydomonas variabilis, Daphnia magna,* and guppies (*Poecilia reticulata*) in different vessels; and/or a simplified artifical ecosystem with sediments, plants, microorganisms, *Daphnia,* and guppies in the same aquarium. *Daphnia* are protected from overpredation in a small case which lets the young crustaceans go out. Simple bioaccumulation and biomagnification in a food chain can both be studied in this system. Accumulation in the sediments can be studied in this system or in a simpler one, without animals.

If the product concentrates in the sediments, further studies of its impacts on the organisms

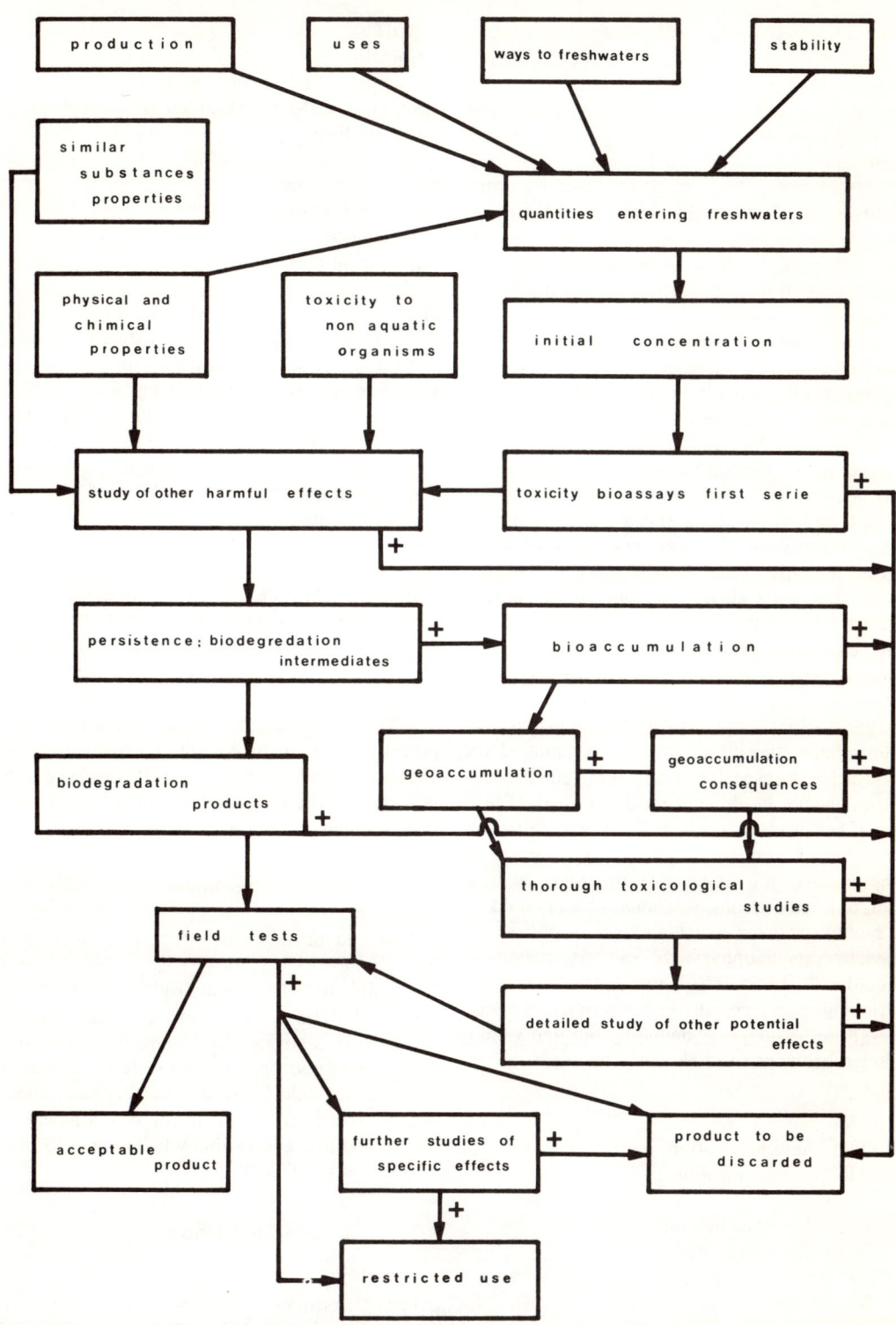

FIGURE 1.—*Scheme for assessing hazard of chemical substance to aquatic life.*

which live in or on the sediments are necessary. Long-term toxicity should be studied in a continuous flow system where an equilibrium between water and sediments can be reached, and at least three species of benthic organisms (insect larvae, a mollusk, and a macrophyte, for example) should be used. If the chemical concentrates in living organisms, a decision to abandon it or to severely restrict its usage has to be made except in some cases of extremely low toxicity to all test organisms, based on a broad coverage of aquatic organisms, birds, and mammals and multiple life cycle testing. As the toxicological work required will be very long and expensive, it is usually better to switch to alternative chemicals if they are available.

Chemicals which do not degrade fast enough or give poorly biodegradable metabolites, but do not accumulate, must be thoroughly tested for long-term toxicity on the species which were the most sensitive in the first series of bioassays. Complete life-cycle and, if possible, multigeneration tests should be done. If the lowest threshold is of the same order of magnitude as the expected environmental conditions, the product should be discarded, or its usage restricted. If the lowest threshold is many times larger than expected critical environmental concentrations, the chemical must be studied for any other foreseeable detrimental effect. This study cannot be done according to a rigid program because the number of possible detrimental effects is very large. All known properties of the product, of its impurities, of its metabolites, and of related chemicals should be examined by a multidisciplinary panel of chemists, physicists, geochemists, biophysicists, sanitary engineers, ecologists, biologists, sewage treatment and drinking water specialists, environmental managers, etc. A checklist including the items mentioned below as examples may be used:

Odor
Odor after chlorination
Odor after ozonization
Taste
Taste after chlorination
Taste after ozonization
Elimination during tap water purification
O_2 transfer through surface
Surface tension of water solution
Interfacial tension of water solution
Foaming
BOD
Chelation
Transportation of other pollutants
Synergism (to be included in toxicological studies)
Micronutrients availability
Eutrophication
Chemoreception
Avoidance
Biological corrosion

This checklist could be expanded by adding all known detrimental effects of existing chemicals and characteristics of aquatic ecosystems that are potentially sensitive to chemicals.

Assays are to be done according to the estimated probability and significance of possible detrimental effects. The critical environmental concentrations should always be used to interpret the results of these assays for making decisions.

If all the laboratory studies seem to show that a new chemical is acceptable, a field test should be done. It can be done either through commercialization in a limited area, or be voluntarily polluting a pond or a small stream, for at least one year, and looking for any ecological (or other) consequence. A continuous input of the chemical is necessary in rivers. It should be proportional to the flow of the river. In ponds, repeated applications can be used.

Conclusion

It would be unrealistic to try to develop a rigid-qualification testing program but it is not impossible to follow a more rigid general scheme with well-defined points of decision. Governments have a strong trend to make compulsory an arbitrary and rigid number of tests because it is easier to build legislation on this basis. Specialists should try to avoid the development of these rigid procedures that will be either too expensive or inefficient.

References

AFNOR (ASSOCIATION FRANÇAISE DE NORMALIZATION). 1974. Determination of the inhibition of the motility of *Daphnia magna* –Strauss 1820. Norme experimentale de l'AFNOR.

AFNOR (Association Française de Normalization). 1977a. Measure of the toxicity to *Brachydanio rerio*. Norme experimentale de l'AFNOR.

AFNOR (Associaton Française de Normalization). 1977b. Measure of the ultimate biodegradability of organic water soluble substances. Norme experimentale de l'AFNOR.

Duthie, J. R. 1977. The imporance of sequential assessment in test programs for estimating hazard to aquatic life. Pages 17–35 *in* F. L. Mayer and J. L. Hamelink, eds. Aquatic toxicology and hazard evaluation. ASTM STP 634. Am. Soc. Test. Mater., Philadelphia.

Lundahl, P. 1974. Determination of the toxicity of *Chlamydomonas variabilis*. Contribution to the study of water pollution by toxic substances. Paris 6 University. 320 pp.

A Sequential Comprehensive Hazard Assessment Strategy for Estimating Water Quality Criteria

J. GARETH PEARSON AND JOHN P. GLENNON

*US Army Medical Bioengineering Research and Development Laboratory
Fort Detrick, Frederick, Maryland 21701*

Abstract

The US Army has been conducting environmental research since the early 1970's for the express purpose of providing to the US Environmental Protection Agency the data-base required for setting water quality criteria for munitions-unique pollutants. In the initial stages of this research program it was recognized that a hazard assessment strategy was required to guide these efforts. What has evolved as a result is a sequential comprehensive strategy for generating the data required for estimating water quality criteria.

The procedure starts with a detailed critical literature review and determination of data gaps, then goes to extensive effluent chemical characterization. Next, sequential research is generally conducted in the following four areas: environmental fate/chemistry; aquatic field surveys; aquatic toxicology; and mammalian toxicology. Research in these individual areas is conducted in sequential phases with discrete decision points at the end of each phase. At these decision points, the environmental data base is assessed and one of three decisions made: (1) the data base is sufficient for recommending a water quality criterion; (2) the data base is insufficient and further research is required; or (3) the data base is sufficient to determine that a water quality criterion is not required.

As research progresses through the research areas, the tasks become more definitive, time consuming, and costly. The environmental fate/chemistry first phase is concerned with making predictive estimates of fate and environmental concentrations while the second phase deals with gathering better estimates of transport and transformation processes. The last phase is concerned with field monitoring under actual or simulated conditions.

The aquatic field studies start with simple diagnostic qualitative estimates of impacts on aquatic communities and can end with definitive evaluations which provide data on the nature and extent of the impacts on aquatic communities.

Both the aquatic and mammalian toxicology studies start with determinations of acute toxicity and terminate with chronic studies which provide estimates of long-term no-effect levels for toxic effects. Results from the mammalian chronic studies are also used in the assessment of carcinogenic potential.

From these data, long-term "safe" pollutant levels are calculated for the protection of man and aquatic organisms. If a compound is found to be carcinogenic, a comprehensive risk assessment must be performed by the appropriate regulatory agency.

The emphasis of this paper is on the aquatic toxicity testing portion of this overall strategy.

The enactment of several items of US Federal legislation during the past decade has precipitated dramatic changes in the attitudes and approaches of scientists and administrators toward improving the quality of this nation's environment. The 1972 amendments to the Federal Water Pollution Control Act mark a particularly significant milestone. That Act established ambitious schedules for the development of water pollutant discharge limitations that would insure the protection of human health and the environment.

As a result of this Act and several more recent items of public health and environmental legislation, concerted efforts have been initiated to improve the capabilities to evaluate the impact of environmental pollutants on man and other life forms. Refinements and standardizations of the philosophies, protocols, and procedures that guide hazard assessments have evolved. In addition, there has been an increasing awareness of the requirement to fully characterize the chemical constituents in pollutant discharges and to define the ultimate fate of these pollutants in the environment. Without such basic information meaningful priorities for the costly and time consuming toxicological evaluations cannot be effectively established.

While modifications to the various tasks that are included in a hazard assessment have improved the quality and comparability of such studies, a generally acceptable management strategy to tie all tasks together has yet to be developed. Several recent papers have described hazard assessment schemes designed to meet specific objectives which serves as a sound philosophical and scientific basis for further development (Duthie 1977; ASTM 1978; AIBS 1978; Kimerle et al. 1978).

The purpose of this paper is to describe a hazard assessment strategy that was specifically developed to evaluate the environmental hazards associated with pollutants unique to the Army's munitions industry and ultimately to recommend pollutant specific water quality criteria to the US Environmental Protection Agency (EPA).

Objective

The overall objective of the hazard assessment strategy is to guide and coordinate the necessary research efforts needed to define the environmental concentration of pollutants that will afford adequate protection to human health and aquatic organisms. The specific objective of the Army's development and use of this hazard assessment strategy was to insure the timely availability of the necessary information from which the EPA and state and local regulatory agencies could formulate environmental quality criteria for pollutants discharged from Army-owned ammunition plants.

Background

The US Army, upon reviewing the 1972 Amendments to the Federal Water Pollution Control Act, recognized that the early availability of environmental criteria for pollutants unique to the munitions industry would improve the cost-effectiveness of pollution abatement programs for the Army's ammunition plants. This would insure that the development of new pollution control technology would be aimed toward the anticipated, rather than presently available, discharge limitations. Thus, pollution control systems designed on the basis of human health and environmental effects data could be expected to have long lifetimes since they would not be rendered obsolete or require extensive upgrading because of the later development of effects data and corresponding discharge limitations.

The US Army Medical Research and Development Command initiated studies in 1972 to define the human health and environmental hazards associated with munitions-unique water pollutants. The program continues to be closely coordinated with the EPA since that agency has the ultimate responsiblity to review the hazard assessment data and to promulgate water quality criteria and effluent limitations. It is emphasized that this research effort was initiated at a time when the EPA was heavily involved in defining the 1977 discharge requirements and had not yet established guidance on the data requirements to achieve 1984 toxicity based discharge requirements.

Scope of Hazard Assessment Strategy

The hazard assessment strategy should be considered a management tool with which to coordinate research efforts and guide decision making rather than a firm protocol that must be adhered to. The strategy defines a sequence in which information should be assembled so that decisions on the priority for continuation of studies of individual pollutants can be justified and documented. There are decision points throughout the strategy to permit review of the assembled information, preliminary estimation of the pollutant's hazard and determination of the requirement and scope of further evaluation tasks. The components of the hazard assessment strategy consist of discrete, but coordinated research tasks and protocols that define how each step of the hazard assessment should be conducted.

It is also emphasized that the hazard assessment strategy being described is not a concept plan. Rather, it is an established approach that has served to guide hazard assessments of Army pollutants for the past 6 years. During this period, many of the individual tasks and protocols within the strategy have been modified or updated to take advantage of improvements and advances in the state of the art in various scientific disciplines. However, the overall strategy has remained relatively unchanged indicating that it is a versatile, effective management tool, even in light of changes in many of the research procedures and protocols.

Although the strategy was developed to define the human health and environment effects of pollutants and to assemble the necessary data to permit the EPA to promulgate water quality criteria, the strategy is well suited for hazard assessments with other specific objectives. For example, this same basic strategy is being used by the Army to conduct hazard assessments of chemicals in industrial manufacturing environments.

The general approach will be briefly presented in order that the aquatic hazard assessment strategy may be viewed in its proper perspective. This will be followed by a detailed description of the aquatic hazard assessment strategy. Description of the relationships between the various protocols is beyond the scope of the paper.

General Approach

The typical sequence of research required for a hazard assessment is shown in Figure 1. The specific tasks that are performed are dependent upon the nature of the compound being assessed and the objective(s) of the assessment.

Problem Definition and Compound Prioritization

Through problem definition studies, information on the effluent source(s) is collected to identify candidate chemical constituents and to document the conditions under which they are discharged to the environment. Data on the characteristics of the receiving system are gathered to define those chemical, physical, and biological properties that will influence the transport and fate of the pollutant and the design of toxicity tests. This is followed by a critical evaluation of the literature to define chemical, physical, and biological properities of the compound and to

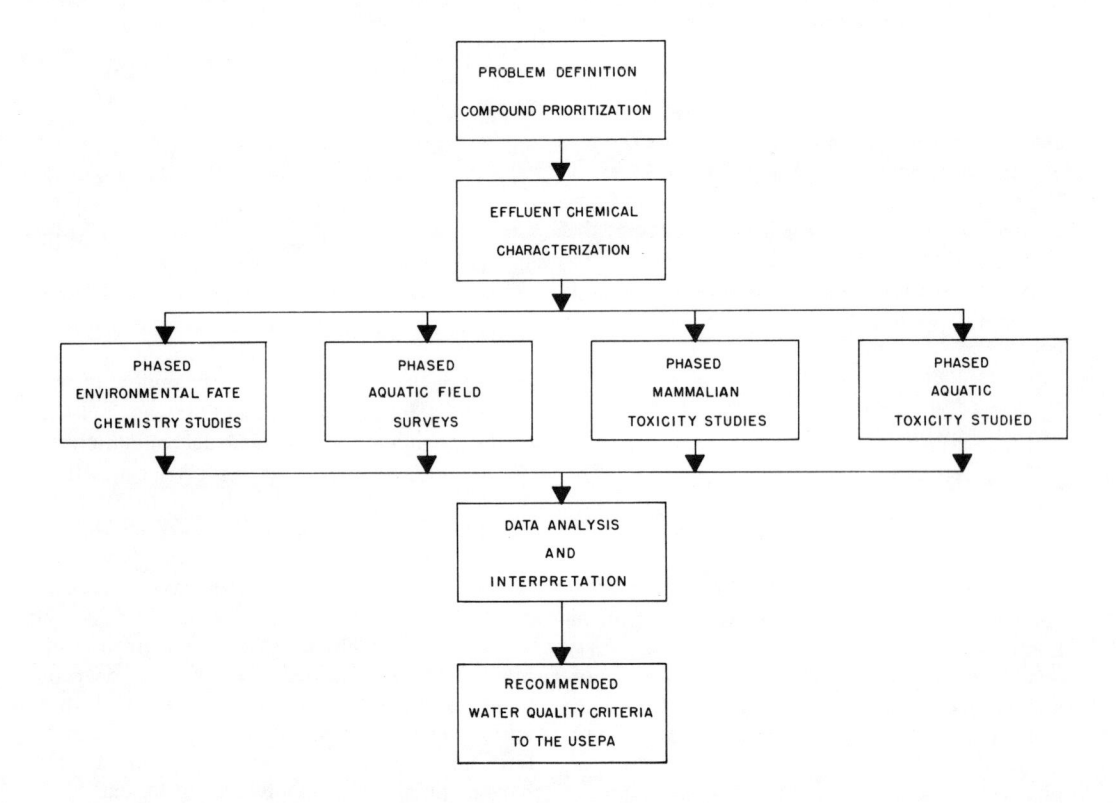

FIGURE 1.—*Typical hazard assessment sequence.*

assess the quality and extent of the available toxicological information. At this stage the data are assembled to identify the munitions-unique compounds, to estimate the extent and magnitude and probably route of exposure to man and aquatic organisms, and to assess the adequacy of the current environmental effects data. If sufficient data are available, water quality criteria will be recommended. However, for munitions-unique water pollutants, sufficient quality information has not been available; therefore, a substantial amount of research has been required to fill the data voids identified by the problem definition study. Table 1 provides additional details as to the major categories covered in a typical problem definition study.

As is the case in most organizations conducting hazard assessments, the number of compounds requiring some level of research far exceed the resources (manpower and economic) available. Therefore, it is essential that some systematic, quantitative scheme of prioritization be employed. This organization has developed, under contract with SRI International, a computerized systems analysis approach for prioritization of compounds on the basis of their estimated hazard. It also allocates resources (dollars) on the basis of reduction in "uncertainly" about the hazard estimate as a result of a specific research task. Hazards are defined in terms of effects predicted by a mathematical model of environmental pollution. An effect can be any adverse biological condition such as cancer in humans, fish kills, or yield loss in crops. To provide a standard comparison among different effects, each effect is assigned a value relative to the others; in the present system, the values are in monetary units and hazards accordingly are expressed in equivalent dollar losses per year.

The fundamental hazard estimate consists of three base elements in the following relationship:

$$H = R \cdot N \cdot V;$$

where H is the hazard; R is the risk of a particular effect; N is the population at risk; and V is the valuation of the particular effect. Risk is a product of the fraction of a given population group that will experience a given effect from a given exposure to a chemical and the estimated concentration in the medium of exposure. Hazard can be aggregated in various ways to yield, for example, the total hazard for a given compound from all locations for a given media, for all population types and groups, and for all effects.

The resource allocation methodology builds upon the hazard estimating system by estimating the uncertainties in aggregate hazards before and after a proposed research study. Uncertainly is defined in the Bayesian subjective sense, in which the scientists who supply the values for the input variables also expresses their subjective evaluation of their accuracy. The uncertainties are then propagated both pre- and poststudy to specific hazards and their aggregations through a Monte Carlo technique. The difference between the pre- and poststudy uncertainties divided by the cost of the research study is used as an index of preference. In practice, many important considerations cannot be treated in this objective allocation methodology. Therefore, the results of the methodology provide guidance but do not dominate the research study selection process. Details of this methodology can be found in Brown (1977) and Small (1977). An example of results from this type analysis for the components of wastewater from the manufacture of TNT can be found in Small (1978).

TABLE 1.—*Major categories of a typical problem definition study.*

Disposal patterns, volumes, and characteristics of effluents and constituents
Physical, chemical, and biological properties of receiving systems
Physical, chemical properties of munitions-unique compounds
Analytical methods
Mammalian toxicology
    Human effects data
    Experimental animal effects data
Environmental considerations
    Environmental fate/chemistry
        Transport
        Transformation
    Environmental effects data
        Wildlife (mammals)
        Birds
        Amphibians
        Reptiles
        Fishes
        Invertebrates
        Microorganisms
        Plants (terrestrial and aquatic)
        Community effects
Existing discharge limitations and standards
Data gaps and voids
Recommended water quality criteria
Recommended research requirements

Effluent Chemical Characterization

Chemical characterization studies are initiated to confirm the identity of individual constituents and to determine the magnitude and consistency of their occurrence. This assumes that sufficiently specific and sensitive analytical chemistry methods are available. If they are not, a substantial analytical method development program must be initiated.

Information from these studies are required to obtain estimates of discharge concentrations and subsequent estimates of expected environmental concentrations. These studies also provide information on the occurrence of production byproducts, unidentified production materials or contaminants, and biological or chemical transformation products from existing pollution abatement systems. For example, in a recent study at a TNT production facility, N-nitrosomorpholine (a known carcinogen) was detected in significant concentrations in the wastewater discharge. Upon further investigation it was found that morpholine was a component of the water treatment chemicals used to treat the production water. When morpholine combined with production wastewater of low pH and high nitrite concentrations N-nitrosomorpholine was probably formed (Spanggord et al. 1978). As a result, the water treatment chemicals have been changed.

The four remaining research areas (see Figure 1) are conducted under the framework of carefully designed research protocols which require data to be generated in sequential phases with decision points at the end of each discrete phase. In all cases, the initial phases are designed to ascertain general properties of the compounds (chemical and/or biological) or gather information on the parameters which might affect their distribution and transformation. Later phases are designed to gather definitive environmental fate/chemistry data and long-term no-effect data for mammals and aquatic organisms.

Environmental Fate/Chemistry

In order to design appropriate toxicological studies upon which a water quality criterion can be based, it is essential to know the form and location of the compound in the environment. Also, if the compound is found to be carcinogenic during laboratory toxicology tests, it is essential that accurate estimates of environmental concentrations, or actual data, be available to faciliate a carcinogenic risk assessment. Since the criteria to be recommended are to be generally applicable to waters of widely differing quality, it is important to determine compound stability under a variety of water quality conditions. This is generally accomplished by determining the effects of varying environmental parameters (e.g., temperature, pH, hardness) on the rate processes which affect the stability of a given compound. It is also important to know the identity of major transformation products early in the testing sequence. The resulting mixture of transformation products along with the major individual pollutants should receive preliminary toxicology assessments in order to determine their relative toxicity. These relative toxicity data may significantly affect the design of subsequent environmental fate/chemistry or toxicology studies, especially if the transformation products are more toxic than the parent compound.

The sequence of testing currently being conducted is depicted in Figure 2. This particular sequence has only been recently formulated and has been derived from reports by Howard (1977), Smith et al. (1977), ASTM (1978), Kimerle et al. (1978), and AIBS (1978).

The testing sequence consists of three phases. The first phase consists of predicting or estimating (from experimentation) the properties and processes which affect transport and transformation of the compound being evaluated. It also includes a comprehensive literature review and tabulation of physical/chemical properties and constants. Table 2 is a partial list of the kinds of data gathered. Experimentation is generally limited to determining relative rates of photolysis, hydrolyses, volatilization, sorption, oxidation/reduction, and biodegradation/transformation. At the conclusion of this phase, preliminary estimates of the fate and expected environmental concentrations are made along with predicted transformation products. Results are then evaluated and requirements for further testing defined.

The second phase is concerned with detailed measurement of properties and processes which affect the transport and transformation of a compound. These studies involve determining rate constants under a variety of conditions and the identification of major transformation prod-

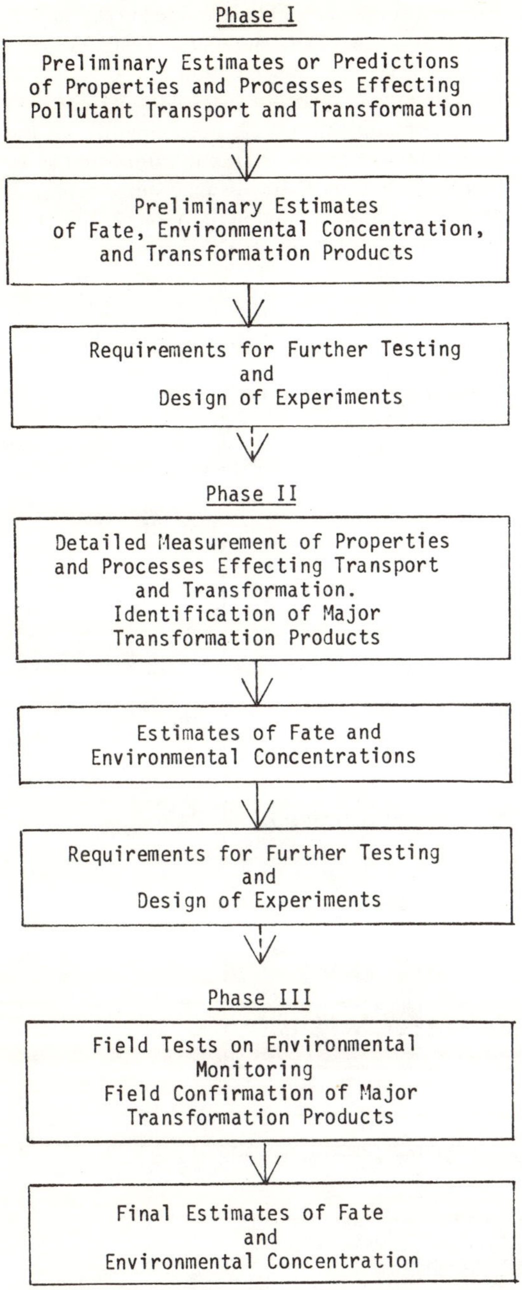

TABLE 2.—*Chemical, physcial, and biological properities and constants required for hazard assessments.*

Water solubility
Gas constant
Vapor pressure
Density
Molecular weight
Structural formula
Boiling point
Melting point
Octanol/water partition coefficients
Sediment/water partition coefficients
UV, visible spectra
Photolysis rate constant
Hydrolysis rate constant
Oxidation rate constant
Biodegradation rate constant
Volatilization rate constant
Complexation rate constants
Sediment desorption

The last phase is concerned with field tests or actual monitoring of aquatic environments for pollutant concentrations in the water, sediments, and biota. These studies can be especially important if results from laboratory tests or predictions are unclear or contradictory.

Aquatic Field Studies

The protocol by which aquatic field studies are conducted was developed in 1973[1] for the Army and recently published by Cairns and Dickson (1978), the original authors.

Effluents from manufacturing plants contain mixtures of waste material and various transformation products of these materials in addition to the production end product. Therefore, it is possible that laboratory-generated toxicity data obtained on single compounds may not adequately define the effects of an effluent on the aquatic environment. Synergistic or antagonistic interactions may cause the actual waste materials to be more or less toxic than would be predicted from laboratory toxicity tests. In addition, it is important to determine whether the effluent causes changes in the aquatic community structure and/or function. Thus, it is desirable to evaluate the impact of the effluent on the aquatic community in the receiv-

FIGURE 2.—*Typical sequence of environmental fate/chemistry testing.*

ucts. At the conclusion of this phase estimates of fate and environmental concentrations are refined and the requirements for further testing again determined.

[1]Cairns, J., Jr., and K. L. Dickson. 1973. Protocol for Evaluating the Effects of Munitions Wastes on Aquatic Life.

ing system through aquatic field surveys. The data base derived from aquatic field surveys is essential information for evaluating the benefits derived from Army pollution abatement programs at the ammunition plants. Figure 3 is a flow diagram of a conceptual aquatic field survey protocol.

Cairns and Dickson (1978) have described the rationale and use of the protocol; therefore, only a limited summary will be provided in this paper. The first phase (Field Screening Tests) consists of two tasks: (1) determination of the effects of the effluent on macroinvertebrate and periphyton communities; and (2) determination of the potential for major effluent constituents to bioconcentrate in fishes or macroinvertebrates. Macroinvertebrates and periphyton are effective study elements since they are substrate-

FIELD SCREENING TEST - Phase I
1. Determine the effects of the effluent on the macro-invertebrate and periphyton communities in the receiving system.
2. Determine if the constituents of the effluent have the potential to undergo biomagnification.

↓

NO MORE TESTS REQUIRED UNLESS INDICATED BY LABORATORY TESTS ← NO / NO — DECISION BOX
1. Is there a significant effect on the macroinvertebrate and periphyton community?
2. Is there biomagnification?
— YES → FOOD CHAIN AND PUBLIC HEALTH STUDIES

↓ YES

EXTENDED FIELD STUDIES - Phase II
1. Determine the effects of the effluent on major elements of the aquatic food web.
2. Determine the effects of the effluent on primary productivity and bacterial activity.

↓

NO MORE TESTS UNLESS INDICATED BY LABORATORY TESTS ← NO / NO — DECISION BOX
1. Is the species diversity and functional capacity of the aquatic community significantly affected by the effluent?
2. Is there an effect on primary productivity or bacterial activity?

↓ YES

DEFINITIVE FIELD STUDY - Phase III
1. Define the extent of response of the affected group of aquatic organisms.
2. Define the extent and conditions under which an effect occurs on primary productivity and bacterial activity.

↓

DECISION BOX
1. Is the effluent having a significant adverse effect on aquatic life?

NO ↓ YES

Data base for evaluating the effects on aquatic life is developed.

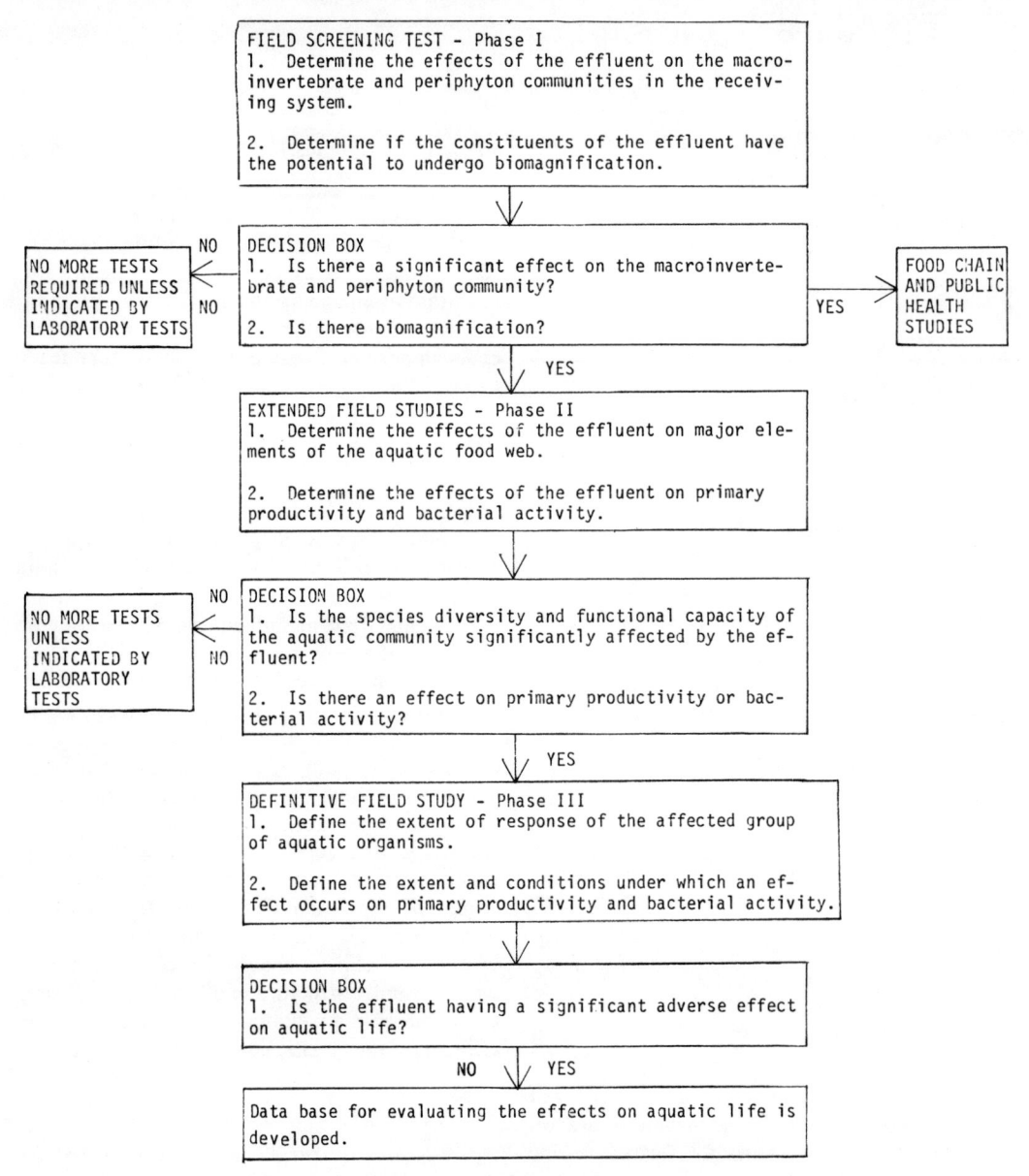

FIGURE 3.—*Protocol for evaluative aquatic field surveys.*

associated and can be easily collected from either natural or artifical substrates. The design of a study must be tailored to the specific waste and receiving system, but must always include reference stations upstream of the discharge and a series of stations appropriately located downstream of the discharge. The potential for bioconcentration is considered because it has public health and wildlife significance. At the conclusion of these studies the data are analyzed and evaluated, and the requirement for additional research documented.

The second phase (Extended Field Studies) also consists of two major tasks: (1) determination of the effects of the effluents on major elements of the food chain; and (2) determination of the effects of the effluent on primary productivity and bacterial activity. These studies include correlations between chemical, physical, and biological elements of the receiving system. The biological portion of the survey should include assessment of the effects of the effluent on fishes, invertebrates, algae, bacteria, and protozoans. The surveys should be carried out under the extremes of environmental conditions in order that the assessment is made under potentially the worst case conditions. To fully evaluate the impact of effluents on aquatic environment, it is necessary to determine effluent effects on primary productivity and the decompositional function of bacterial communities. At the conclusion of this phase the data are also analyzed and evaluated and the requirements for additional research defined.

The third phase (Definitive Field Studies) is concerned with further defining the nature and extent of effects on aquatic organisms, primary productivity, and bacterial activity. Methods cannot be explicitly defined since they will depend on the unique set of conditions and effects observed in the receiving system during previous surveys. At the conclusion of these studies sufficient information should be available to determine the nature and extent of effects on the biota of the receiving system. These data are then integrated into the entire hazard assessment strategy and influence the subsequent derivation of appropriate water quality criteria. Examples of the use of this protocol in evaluating the effects of discharges from Army ammunition plants can be found in Stilwell et al. (1976), Sanocki et al. (1976), Sullivan et al. (1977), and Sullivan et al. (1978).

Toxicity Studies

Toxicity studies (both aquatic and mammalian) are conducted on individual pollutants or compound mixtures which are representative of actual pollution discharges. Single-compound testing has the advantage that a specific dose-response relationship can be established. Testing of compound mixtures has the advantage of identifying compound interactive (synergistic, antagonistic, or additive) effects that might result from exposure to actual discharges. While both types of testing (single-compound and mixture) may be necessary to develop a complete toxicological data base, this is not always practical or economical.

Some toxicity studies designed to establish environmental criteria for complex waste mixtures have not included definitive studies on those mixtures. In those instances, preliminary evaluations indicated the toxic properties of the waste mixture were dominated by its principal component(s). An example is the mammalian toxicological evaluation of wastewaters from nitroglycerin production. Chemical characterization of Radford Army Ammunition Plant wastewaters confirmed the principal munitions-unique component to be nitroglycerin. Appreciable quantities of lesser nitrated esters of glycerol (glyceryl-1, 2-dinitrate [DNG], 1, 3-DNG, glyceryl-1-nitrate [MNG], and 2-MNG) also were detected in these effluents. All five compounds underwent acute (LD50) testing. Results indicated that all isomers have approximately equal toxic properties (Lee et al. 1975; Lee et al. 1977). Thus, it is unlikely that the toxicity of the wastewater mixture would be dominated by a minor component. In addition, metabolic studies indicated that nitroglycerin is sequentially transformed to the lesser nitrated esters of glycerol during animal metabolism (Lee et al. 1977). Since exposure of laboratory animals to nitroglycerin also would involve metabolic exposure to the other esters, plans to evaluate compound interaction effects in separate studies were deleted.

In the Army's environmental criteria research program the current studies with condensate water (TNT production wastewaters) involve testing of a complex mixture. Chemical characterization studies of wastewater from Joliet and Volunteer Army Ammunition plants indicated that these effluents contained in excess of 30 munitions-related organic compounds. Detailed single-compound testing of such a complex mixture would not be economically practical. In addition, single-compound testing would fail to evaluate compound interactive effects. Thus, a strategy was formulated to develop the necessary data base on a complex industrial wastewater (Pearson et al. 1979). This strategy involves three general tasks:

1. Conduct detailed toxicological studies (mammalian and aquatic) on the "representative" mixture of the wastewater.
2. Conduct screening toxicological evaluations (mutagenic screening and acute aquatic toxicity tests) of all identified wastewater components. This will screen compounds for extraordinary toxic properties and influence their priority for inclusion in the "representative test mixture."
3. Conduct detailed toxicological studies on selected wastewater components. Priority for this testing is based on the compound's relative concentration in the wastewater and its relative toxicity as determined above.

The "no-effect" level defined during toxicity studies with the mixture will be the basis for an environmental criterion. This will be supplemented with the identification of extraordinarily toxic wastewater components and the definition of the dose-response relationship for principal compounds in the mixture.

Mammalian Toxicity Studies

A general protocol was developed to determine the potential toxicity of environmental pollutants to man using laboratory animals. The protocol generally conforms to those procedures detailed in World Heath Organization (1969), Sontag et al. (1976), National Research Council (1974), and National Academy of Sciences (1977). The protocol is divided into three phases: (1) effects of acute exposure; (2) effects of multiple exposure; and (3) effects of lifetime exposure. This sequence is depicted in Figure 4.

The objectives of the first phase (Acute Toxicity Studies) are to establish a dose-response relationship between the test animal and test substance, to identify target organs involved, to establish appropriate species for further studies

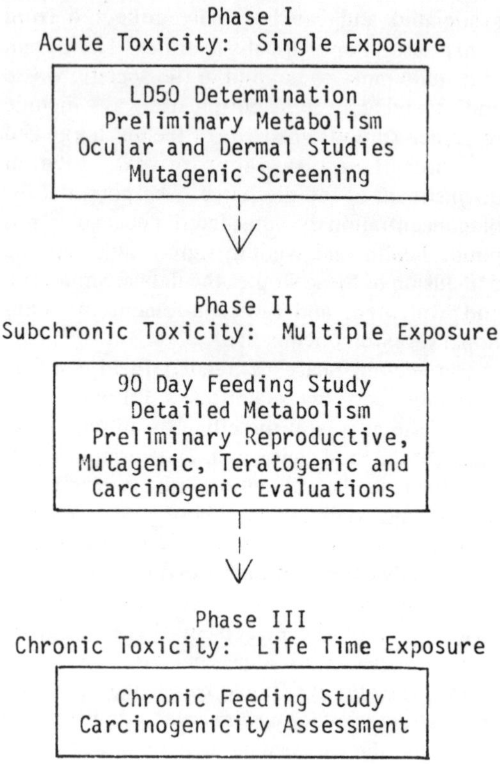

Phase I
Acute Toxicity: Single Exposure

LD50 Determination
Preliminary Metabolism
Ocular and Dermal Studies
Mutagenic Screening

Phase II
Subchronic Toxicity: Multiple Exposure

90 Day Feeding Study
Detailed Metabolism
Preliminary Reproductive,
Mutagenic, Teratogenic and
Carcinogenic Evaluations

Phase III
Chronic Toxicity: Life Time Exposure

Chronic Feeding Study
Carcinogenicity Assessment

FIGURE 4.—*Schematic of the mammalian toxicity protocol.*

and to identify the most sensitive species. Median lethal doses (LD50) are determined for male and female rats and mice, and possible other test species. Radioactively labeled test compounds are generally used for preliminary estimates of adsorption, distribution, excretion, biotransformation, and pharmacokinetics. Dermal and ocular studies are conducted according to methods recommended by the Food and Drug Administration.[2] Mutagenic screening tests are also conducted according to the methods of Ames (Ames et al. 1972; Ames et al. 1973a; Ames et al. 1973b; McCann et al. 1975; Ames et al. 1975).

The second phase is concerned with subchronic, multiple-exposure type studies. The objective of these studies is to further define the nature and extent of the interactions between the test substance(s) and selected biological systems

---

[2]Federal Register 16 CFR 1500.41 91 26, 141, and Federal Register 16 CFR 1500.42 91 26, 142.

by identification of the lesions that could lead to pathological manifestations at the cellular and other levels, and their reversibility. These studies are also designed to further elucidate the dose-response relationship as a result of multiple dosing.

Subchronic studies lasting 90 days are conducted with mice and rats, in addition to 3-6-month studies with dogs. Definitive studies of absorption, distribution, excretion, biotransformation, and pharmacokinetics are conducted. Preliminary reproductive and teratogenic evaluations are made. Any further mutagenic-carcinogenic screening, involving *in vitro* systems, are also completed during this phase.

The effects of lifetime exposure (phase three) are usually determined on rats and mice; however, other animal species are used when appropriate. These studies further define the nature and extent of the interaction between the test substance(s) and the selected biological systems. This definition includes identification and/or verification and specificity of the lesion(s) at the biochemical and cellular level. These studies are also designed such that the compound(s) can be assessed for their carcinogenic potential. Necessary reproductive, teratogenic, and mutagenic assessments not completed during the first two phases are completed during this phase.

Results from these studies can cause a compound to be classified carcinogenic to laboratory animals. Such a determination has a significant impact on how the data are analyzed to arrive at a recommended water quality criterion for the protection of man. The exact policy and procedures in this case have yet to be defined by the EPA. If, however, a compound is not found to be carcinogenic in laboratory animals, a recommended water quality criterion can be derived by relatively simple mathematical procedures. The procedures used to evaluate mammalian toxicity data for noncarcinogenic compounds were developed by the World Health Organization (1974), and have been used extensively by that organization to develop Acceptable Daily Intake (ADI) values. This value is then converted to an acceptable drinking water level through a simple transformation assuming the "standard" human uses 2 liters of water per day for dietary purposes. Details of these conversions as they are used by the Army can be found in later sections of this paper.

Aquatic Toxicity Protocol

The protocol by which aquatic toxicity tests are conducted was developed for the Army in 1973[3] and recently published by the same authors (Cairns and Dickson 1978). Since one of the objectives of this paper is to illustrate the use of the Army's strategy for assessing the effects of conventional munitions on aquatic organisms, this section shall be discussed in greater detail than the others.

The scope of this protocol was limited to freshwater organisms since none of the compounds of Army interest are released into or near estuarine or marine environments. With minor modification, the protocol could easily be expanded to assess the potential hazard of pollutants in other habitats.

The protocol is divided into three phases: (1) acute static toxicity tests; (2) acute flow-through toxicity tests; and (3) chronic toxicity tests (Figure 5). Progression from one phase to another is dependent upon the results from within this protocol and from those just presented. At intermediate milestones (these usually coincide with the end of each of the first two phases of the protocol) the data base is assessed and one of the three decisions made. These are : (1) the data base is sufficient for recommending a water quality criterion; (2) the data base is insufficient for recommending a water quality criterion and further research is required; or (3) the available data base indicates that a water quality criterion is not required.

Phase I–Accute Static Toxicity Tests

The purpose of the first phase is to obtain preliminary estimates of the effect of a chemical on aquatic life and to determine the need for a water quality criterion for the protection of aquatic organisms for a given chemical. It is comprised of five tasks: (1) relative sensitivity of fishes, invertebrates, and algae; (2) relative sensitivity of various fish life stages; (3) effects of temperature, pH, and hardness on the toxicity to fish; (4) sta-

---

[3]Cairns, J., Jr., and K. L. Dickson. 1973. Protocol for Evaluating the Effects of Munitions Wastes on Aquatic Life.

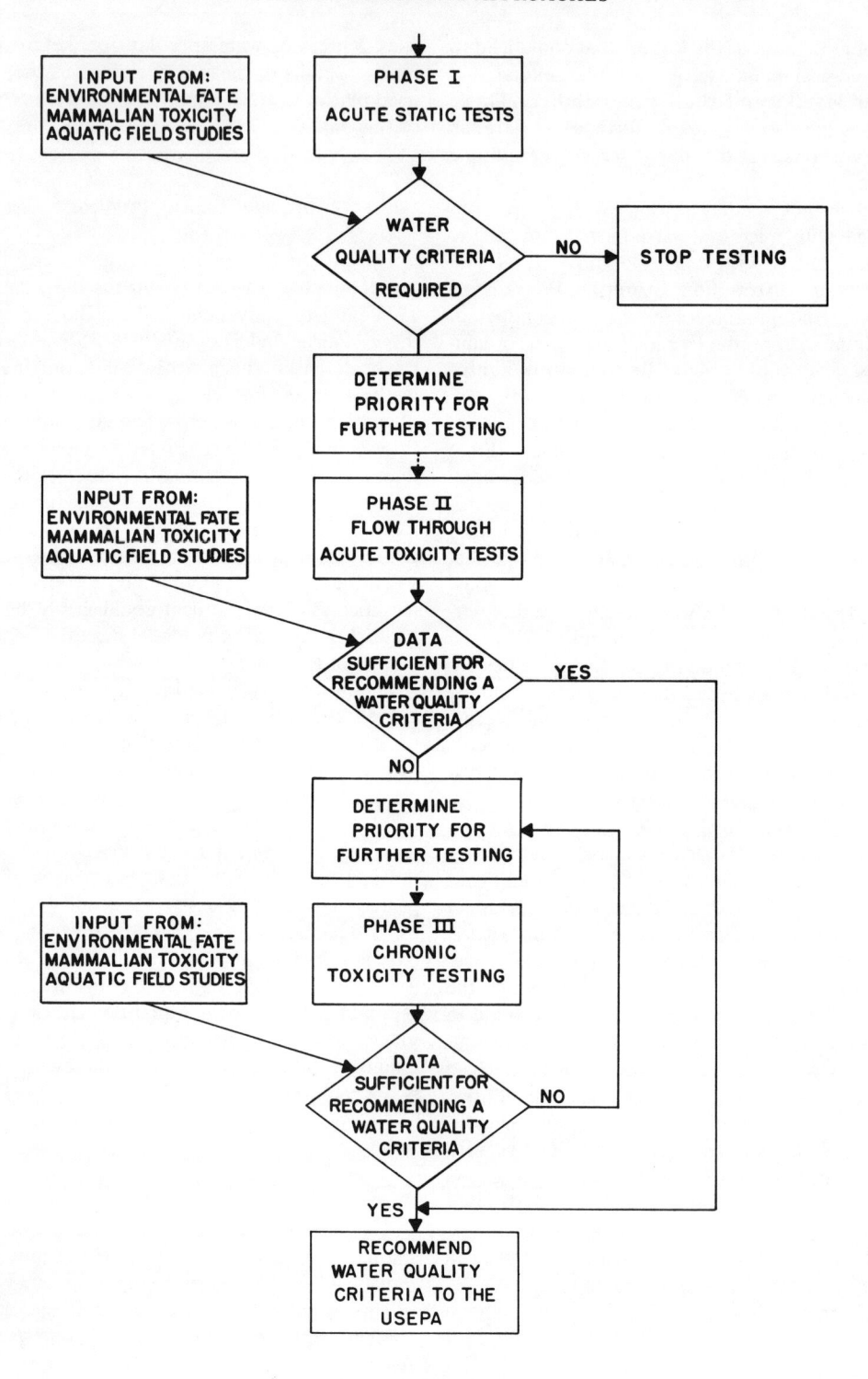

FIGURE 5.—*Schematic of the aquatic toxicity protocol.*

bility of the toxicological characteristics of the toxicant; and (5) potential for fish flesh tainting and bioaccumulation. All toxicity tests conducted during this phase employ static procedures over a wide range of nominal concentrations.

Task 1

The relative sensitivity of aquatic organisms is defined by determining median effect levels of four species each of fishes, invertebrates, and algae. Also, no-effect levels for algal species are determined by using a one-way analysis of variance and an appropriate multiple comparison technique, if indicated. The static acute toxicity tests on fishes and invertebrates are conducted according to the methods described by Stephan (1975). Algal toxicity tests are patterned after methods described by the USEPA (1971). Those species which have been used in the Army's program are listed below:

Fishes
Bluegill	*Lepomis macrochirus*
Rainbow trout	*Salmo gairdneri*
Channel catfish	*Ictalurus punctatus*
Fathead minnow	*Pimephales promelas*

Invertebrates
Water flea	*Daphnia magna*
Amhipod	*Gammarus fasciatus*
Isopod	*Asellus militaris*
Midge larvae	*Chironomus tentans*
Midge larvae	*Tanytarsus dissimilis*
Oligochaete	*Lumbriculus variegatus*
Amphipod	*Hyalella azteca*

Algae
Cyanophyte	*Microcystis aeruginosa*
Cyanophyte	*Anabaena flos-aquae*
Chlorophyte	*Selenastrum capricornutum*
Chrysophyte	*Navicula pelliculosa*

Fish toxicity tests will rarely suffice as long as the possibility exists that the organisms upon which the fish depend may be eliminated by concentrations of the toxicant tolerated by fish, which would then indirectly affect the fish population. Therefore, for the purpose of estimating water quality criteria it is essential that, at a minimum, the toxicity to representatives of the above elements of the food chain be determined. The most sensitive group as well as species within a group should be determined since this will influence the design of later research. Both inhibitory and stimulatory effects on algae should be evaluated since both cause significant changes in aquatic communities.

Task 2

The objective of this task is to determine the relative sensitivity of various fish life stages to a particular compound. Here again, median lethal levels are determined under static test conditions. Generally, LC50's are determined for fathead minnow eggs, 1-hour posthatch larvae, 7-day posthatch fry, 30-day posthatch juveniles, and 60-day posthatch minnows. These tests provide information on the potential susceptibility of critical life stages and aid in the design of further research.

Task 3

Changes in water quality may alter the susceptibility of aquatic organisms either through changes in the chemical species or in the aquatic organism. The effects of pH on the toxicity of free cyanide to fathead minnows is an excellent example (Broderius and Smith 1977). Also, when developing a data base to support a water quality criterion, it is essential that the applicability of this criterion to the Nation's various water qualities be known. Obviously, extensive environmental fate data will provide some of the answers about changes in the chemical, but it is also necessary to obtain data on the toxicological consequences of these changes. This can be accomplished by determining the acute toxicity of a compound to a sensitive species under various water quality conditions. These generally are pH (6, 7, 8), temperature (15, 20, 25°C), and hardness (35, 100, 250 mg/liter as $CaCO_3$), but additional parameters may also be evaluated depending on the chemical nature of the compound being considered. If these parameters produce markedly different effects on the species being tested, additional tests should be conducted to further define the degree of variability likely to result from even greater changes in these parameters. In addition, results of this nature will also influence the direction and intensity of environmental fate studies and may change the priority for determining the effects of various processes (e.g., hydrolysis, complexation) on the compound.

Task 4

It is also important in this phase to determine whether any transformation products of the parent compound are more or less toxic than the parent compound. Preliminary estimates of this are more or less toxic than the parent compound. Preliminary estimates of this are obtained by determining the stability of the toxicological characteristics of the compound by conducting tests in which the test solutions have "aged" 12, 24, 48, and 96 hours before initiating the toxicity test. Results from these tests will provide some basis for establishing the priority for qualitative and quantitative identification of transformation products within the environmental fate/chemistry protocol.

Task 5

The purpose of the final task within the first phase is to determine the potential for fish flesh tainting and bioconcentration. The potential for fish flesh tainting is especially important since most water quality criteria will be applied to waters designated for recreational or commercial uses. Tainted fish flesh can potentially render a fisheries resource unusable, resulting in significant economic impact. It has been our experience, however, that it is not generally possible to get fish flesh tainting tests conducted unless a substantial amount is known about the human health effects of the compound being evaluated. For all of the compounds we have had experience with, sufficient human effects data have not been available and, therefore, no fish flesh tainting studies have been conducted. It is also important to obtain information on the potential for a compound to bioconcentrate. Recently, factors like water solubility and octanol-water partition coefficients have been shown to be useful for predicting bioconcentration potential (Neely et al. 1974). In those situations where the predicated bioconcentration factor (BCF) falls into a "gray area" (e.g., BCF between 800 and 1,500) it may be appropriate to conduct limited bioconcentration tests with radioactively labeled material. These results can then be used to determine the need for detailed bioconcentration tests during the next phase.

At the completion of the first phase the data are assessed and the requirements for a water quality criterion are determined. The decision criteria for recommending that a water quality criterion *not* be developed are : (1) the compound is relatively non-toxic (LC50's and EC50's $\leq$ 1,000 mg/liter); (2) the compound's toxicity is not increased by environmental parameters or "aging"; (3) the compound has a low potential for bioconcentrating (BCF $\leq$ 100); and (4) results from mammalian and environmental fate/chemistry studies also indicate that the compound has a low potential for causing adverse environmental effects. If any of these criteria are not met, some level of testing will be required. The complexity of the follow-on studies will vary according to the relative hazard of the compound as predicted from this first phase. Obviously, compounds which are highly toxic will require extensive evaluations to insure that the impact of their presence in the environment is well understood. For moderately toxic compounds, additional studies should be conducted to clarify any areas of uncertainty. In all cases where further research is required it is essential that these studies be designed on the basis of the most recent information. Further, it is recommended that the priority for these studies be determined using those techniques described in the section on compound prioritization.

Phase II – Acute Flow-Through Toxicity Tests

The second phase of the protocol is designed to obtain more definitive data on the acute toxicity and initial estimates of the chronic toxicity. Most tests are conducted under flow-through conditions with toxicant concentrations analytically measured. Phase II consists of five tasks: (1) relative sensitivity of aquatic organisms; (2) embryo-larval toxicity tests; (3) detailed bioconcentration studies; (4) further investigation of the effects of environmental parameters on toxicity; and (5) evaluation of the toxicological properties of transformation products or mixtures (see Pearson et al. 1979) for additional details on this task).

Task 1

The relative sensitivity of aquatic organisms is determined on four species of fish and invertebrates according to methods described by Stephan (1975). The species tested should be, at a minimum, the most sensitive as determined during Phase I. If the relative sensitivity (LC50's or EC50's) of the fishes and invertebrates from Phase I has a narrow range, the number of species tested could be reduced. Median-effect levels (LC50's or EC50's) are again determined and compared to those from Phase I. Time-independent, median-effect levels (incipient LC50's) are also determined. Evaluation of these data should provide evidence as to whether a compound has the potential to be cumulatively toxic.

Task 2

Embryo-larval toxicity tests are conducted on two species (generally fathead minnows and channel catfish) according to methods described by Macek and Sleight (1977) and McKim (1977). Similar tests on rainbow trout are conducted if it has been shown that the trout is a much more sensitive species or that the toxicant is known or projected to be released into cold water habitats (USEPA 1972a). This test is more expensive than the other embryo-larval tests because of the longer embryo-larval development time required for rainbow trout. Therefore, its use must be entered into carefully and be strongly justified. Studies with fathead minnows and channel catfish are initiated with recently fertilized eggs and routinely continued for 30 days. These tests, however, are designed such that they could be carried into full chronic studies. The decision to do this will depend on study requirements and whether or not significant effects have been observed in the two highest test concentrations by day 30, the test should be stopped and lower concentrations selected for any subsequent chronic tests. Macek and Sleight (1977) have shown that for many compounds the estimated maximum acceptable toxicant concentration (MATC)[4] from the embryo-larval test is a reasonable approximation of the MATC from full life-cycle tests on the same compound and species. However, since these are relatively short tests, the estimated MATC should be used with caution especially if there is evidence that the compound is cumulatively toxic.

Task 3

The objective of conducting detailed bioconcentration tests is to obtain an experimental derived estimate of the bioconcentration factor (BCF) and to determine relative estimates of the uptake and depuration rates. Tests are continued until the net uptake rate over time is not statistically different from zero (i.e., a plateau has been reached). The test animals are then transferred to clean containers with a continuous flow of uncontaminated water and monitored until whole-body residues have reached at least 25% of the plateau value. It is generally preferred to monitor the chemical concentration in both the test water and organism by direct analytical determination. However, if a suitable analytical method is not available it may be necessary to use radioactively labeled material. At the conclusion of these tests, uptake and depuration rates are determined along with the bioconcentration factor. Use of a kinetic approach to bioconcentration tests (Branson et al. 1975) may significantly reduce the time and cost of these tests while still providing accurate data.

Task 4

This task is to further investigate the effects of environmental parameters on a compound's toxicity. The specific design will depend on the results from Phase I or perhaps from data generated during environmental fate testing.

At the conclusion of Phase II, the data are assessed and estimates made of the compound's relative hazard. If the purpose for conducting some Phase II studies was to clarify "gray area"

---

[4] MATC is the maximum concentration that causes no statistically significant effects on the reproduction, growth, and survival on an organism during a life-cycle test (Mount and Stephan 1967).

encountered during Phase I, the determination should be made as to whether a water quality criterion is required. If a criterion is not required, testing should be terminated. If a criterion is required, additional Phase II or Phase III studies should be designed. The priority for conducting further studies on a particular compound should also be determined.

If the purpose of the Phase II studies was to obtain preliminary estimates of a water quality criterion, then the data-base should be assessed and a water quality criterion calculated, if appropriate. If a compound is only moderately toxic (LC50 or EC50 > 10 mg/liter), there is not evidence of cummulative toxicity, the BCF is less than 100, depuration is rapid, and data from the other test protocols are supportive, then a water quality criterion can be recommended on the basis of the MATC value from embryo-larval toxicity tests. If data from the other protocols are not supportive or if any of the above criteria are not met, testing should proceed to Phase III and/or some of the Phase II studies should be redesigned.

Phase III—Chronic Toxicity Tests

Phase III studies consist of life-cycle tests on fathead minnows (USEPA 1972b), *Daphnia magna,* (Biesinger 1975) and a larval midge (Bentley et al. 1978). If required, life-cycle or partial life-cycle tests may also be conducted on rainbow trout (USEPA 1972a). All tests are conducted under flow-through conditions with toxicant concentrations analytically measured in all test containers at predetermined times. Life-cycle tests with fathead minnows are initiated with recently fertilized eggs or newly hatched fry and are continued through 30 days posthatch of the F_1 generation. Survival, length, and weight are determined at intermediate times (30, 60, and ~175 days) on the F_0 generation and at 30 days posthatch on the F_1 generation. In addition, the following parameters are determined at termination of the F_0 generation for each control and experimental treatment: length and weight (both sexes); total spawns; total eggs; spawn/female; eggs/female; and eggs/spawn. Effect and no-effect concentrations for each parameter are determined using analysis of variance techniques and multiple comparison tests. The ratio of the highest no-effect concentration (MATC) from this test and the lowest 96-hour LC50 for the fathead minnow is calcultated as the fish application factor (AF) (Mount and Stephan 1967). Use of the AF will be discussed in the next section.

Life-cycle tests with *Daphnia magna* are initiated with animals less than 24 hours old and continued for 21 days. The F_1 generation is also continued for 21 days. Survival and the number of young per parthenogenetic female are determinated and the data analyzed in a manner similar to the fathead minnow chronic data. 'The MATC and AF are also determined for *Daphnia magna*. For details see Biesinger (1975) or Bentley et al. (1978).

For the midge larvae life-cycle test, animals less than 48 hours old are used and the F_0 is continued through emergence and the deposition of eggs. The F_1 is conducted in a similar manner. For each generation the following parameters are determined: survival of larvae, pupae, and adults; percent emergence; and the number of eggs per adult. These data are also analyzed in a manner similar to the fathead minnow chronic data with the MATC and AF determined. For details see Bentley et al. (1978).

Calculation of the Water Quality Criteria

Within the Army's current hazard assessment program, two criteria are recommended for each compound: one for the protection of human health; and one for the protection of aquatic organisms.

Water Quality Criteria for the Protection of Human Health

Results from toxicity studies with mammalian species can cause a compound to be classified as carcinogenic to laboratory animals. This determination has a significant impact on how the data are further analyzed to arrive at recommendations for water quality criteria. If the compound is found to be carcinogenic in laboratory animals, the data are subjected to a complex risk assessment procedure. Data from compounds not found to be carcinogenic in laboratory animals can be evaluated by relatively simple mathematical procedures. In either case, a conservative approach is adopted because the toxicity studies are conducted with laboratory animals which are not the target organism of principal concern, i.e., human. The decision as to which procedure is appropriate usually is not possible until chronic studies have been completed. This is because chronic studies are the only studies where definitive evaluations for carcinogenicity are made.

A detailed description of the risk assessment procedure is beyond the scope of this paper. In essence, however, it utilizes computer modeling techniques to predict human cancer rates associated with various human exposure (environmental) levels. The environmental discharge of the compound should then be controlled to a level corresponding to the maximum acceptable human cancer rate. While this approach appears relatively simple, lack of universal acceptance in several areas has prevented its full implementation. The principal area of uncertainly is a definition, by the regulatory agencies and/or the scientific community, of what is an acceptable human cancer rate for environmental pollutants. The Food and Drug Administration (FDA), in establishing criteria and procedures for evaluating assays for carcinogenic residues in food producing animals, has stated that a one-in-one-million human cancer rate is an acceptable maximum risk level (Federal Register 1977). FDA maintains that such a risk level can be considered an insignificant public health concern. This is due to the many conservative requirements inherent in the extrapolation from animal data to the human and because the estimated risk is extremely low relative to all other hazards to the human population, e.g., other diseases or accidents. The EPA has not yet proposed a uniform acceptable risk level for environmental pollutants found to be carcinogenic. EPA presently evaluates carcinogenic pollutants on a case-by-case basis with the Administrator having the responsibility for defining an acceptable human cancer rate for each compound.

The procedures used to evaluate mammalian toxicity data for noncarcinogenic compounds was developed by the World Health Organization (1974) and has been used extensively by that organization to develop Acceptable Daily Intake (ADI) values. The basis for such calculations is the highest dose level which causes no adverse toxicological effects (no-effect dose level during animal studies). Ideally, this no-effect level is based on long-term (chronic) studies. When such data are not available the highest no-effect dose is selected from subchronic studies. In studies where several species have been evaluated, data from the species showing greatest sensitivity to the test compound are used. The World Health Organization (1974) also recommends that extrapolation of animal data to humans include the use of a safety factor. They cite the following reasons for its use: "to allow for any differences in sensitivity between the animal species and man; to allow for wide variations in sensitivity among the human population; to allow for the fact that the number of animals tested is small compared with the size of the human population that may be exposed." A safety factor of 100 is applied to reliable data from chronic studies while a safety factor of 1,000 is used when only partial chronic or subchronic testing has been completed. All calculations in the Army's program are for water pollutants where the principal route of human exposure is assumed to be from dietary water, e.g., use of contaminated surface water for municipal or private drinking water supplies. Under these conditions a water quality criterion for the protection of human health is calculated in the following manner:

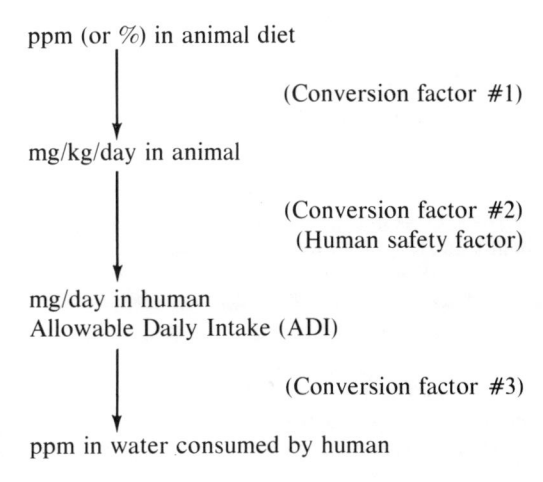

ppm (or %) in animal diet
$\downarrow$ (Conversion factor #1)
mg/kg/day in animal
$\downarrow$ (Conversion factor #2)
   (Human safety factor)
mg/day in human
Allowable Daily Intake (ADI)
$\downarrow$ (Conversion factor #3)
ppm in water consumed by human

Conversion Factor #1

The calculation is required only when animal data are reported in terms of parts per million (or percent) of the test compound (TC) in the animal feed. Conversion to mg/kg/day requires knowledge of animal daily food consumption (DFC) and animal weight (AW) and is accomplished by the following equation:

$$\text{mg/kg/day (in animal)} = \frac{\text{ppm(TC)} \times \text{g(DFC)}}{\text{g(AW)}}$$

or

$$= \frac{\%(\text{TC}) \times 10^4 \text{g(DFC)}}{\text{g(AW)}}.$$

Conversion Factor #2

This calculation transforms the animal daily dose level (in terms of body weight) to the daily total dose for a 60-kg "standard" human.

Human Safety Factor

At this time, the 100 or 1,000 safety factor is also applied to account for the uncertainty in animal data to predict human effects. The overall equation for both calculations is:

$$\text{mg/day (in "standard" human)} = 0.01 \text{ (or } 0.001) \times 60 \times \text{mg/kg/day (in animal)}.$$

This is the ADI or the maximum dose level to insure protection of human health.

Conversion Factor #3

This calculation is specific for the route of human exposure (dietary water) and is required to permit expression of the acceptable human dose in terms of a measurable pollutant concentration. In this case, where the "standard" human consumes an average 2 liters of water per day, the calculation is:

$$\text{ppm (in drinking water)} = \frac{\text{mg/kg in "standard" human}}{2 \text{ kg water/day}}$$

In this relatively simple situation, human exposures from sources other than drinking (dietary) water (atmosphere, food, etc.) are not considered. If studies indicate there is a significant potential for multiple routes of exposure, e.g., determination of significant bioconcentration by aquatic organisms in the human food chain, summation of the human dose from all such exposures should be made during this final calculation. Additonal guidance for these calculations has been published by the Food and Drug Administration (1959).

Water Quality Criteria for the Protection of Aquatic Organisms

Two methods have been used by the Army to calculate water criteria for the protection of aquatic organisms. The first method involves the use of an experimentally derived application factor (AF) as described in the previous section. This fractional value has been shown to be constant for many compounds across broad taxonomic groups of aquatic organisms, and therefore can be utilized to estimate chronic safe levels based on acute toxicity data (Mount and Stephan 1967; Macek et al. 1976a; Macek et al. 1976b). For all data sets, the AF is multiplied by the lowest 96-hour LC50 for the same species, and the lowest resultant value selected as the water quality criterion. This methodology has recently been reviewed and found not to be appropriate for some compounds (Mount 1977; Andrew et al. in press). As an alternative method, when the application factor is considered inappropriate, a somewhat arbitrary safety factor is applied to the highest no-effect value from a partial or full life-cycle toxicity test. This product is then the water quality criterion. Recently, the EPA has published a proposed method for calculating water quality criteria (Jorling 1978). An initial assessment of that methodology with respect to data on munitions compounds indicated that the water quality criteria derived by either of the above procedures do not differ significantly from those calculated from the proposed procedure.

Conclusions

The strategy used by the Army has proved to be useful in developing the data required for recommending water quality criteria for munitions-unique pollutants. The use of a centralized manager/management scheme over all research activities facilitates the integration and coordination of the diverse efforts, avoids duplication, and insures the timely availability of the required data.

It should be stressed that no scheme or numerical criteria for decision making will replace the need for good scientific judgment. This is especially true in the later phase of this strategy where a multitude of facts must be integrated from numerous scientific disciplines into a final hazard assessment or water quality criterion.

As testing proceeds through the phases, the "true" estimate of a pollutant hazard becomes better. In addition, our confidence (decrease in uncertainty) in that estimate also increases. Efforts to quantify the uncertainly about a hazard estimate should be pursued in order to allow a better understanding of the risks one might be willing to take with various degrees of knowledge. Brown (1977) has made significant advances in this area and they should be continued.

While the uncertainty associated with various levels of knowledge (testing) about a compound is quantifiable, the "acceptable risk" associated with a specified level of a pollutant released to the environment is basically a sociopolitical judgment and not subject to the rigors of scientific investigation.

Acknowledgments

It is not possible to acknowledge all those persons who have contributed to the present form of this hazard assessment strategy. Suffice it to say that we are truly grateful to all of those who have contributed their time, intellect, and, on occasion, indulgence.

References

AIBS (AMERICAN INSTITUTE OF BIOLOGICAL SCIENCES). 1978. Criteria and rationale for decision making in aquatic hazard evaluation. Pages 241–273 in J. Cairns, Jr., K. L. Dickson, and A. W. Maki, eds. Estimating the hazard of chemical substances to aquatic life. ASTM STP 657. Am. Soc. Test. Mater., Philadelphia.

AMES, B. N., W. E. DURSTON, E. YAMASAKI, AND F. D. LEE. 1973a. Carcinogens are mutagens: a simple test system combining liver homogenates for activation and bacteria for detection. Proc. Nat. Acad. Sci. U.S.A. 70:2281–2285.

AMES, B. N., E. G. GURNEY, J. A. MILLER, AND H. BARTSCH. 1972. Carcinogens as frame-shift mutagens: metabolites and derivatives of 2-acetylaminofluorene and other aromatic amine carcinogens. Proc. Nat. Acad. Sci. U.S.A. 69:3128–3132.

AMES, B. N., F. D. LEE, AND W. E. DURSTON. 1973b. An improved bacterial test system for the detection and clarification of mutagens and carcinogens. Proc. Nat. Acad. Sci. U.S.A. 70:782–786.

AMES, B. N., J. McCANN, AND E. YAMASAKI. 1975. Methods for detecting carcinogens and mutagens with the *Salmonella*/mammalian-microsome mutagenicity test. Mutat. Res. 31:347–364.

ANDREW, R. W., D. A. BENOIT, J. G. EATON, J. M. McKIM, AND C. E. STEPHAN. (In press.) Evaluation of an application factor hypothesis. U.S. Environ. Prot. Agency Ecol. Res. Ser., Duluth, Minn.

ASTM (AMERICAN SOCIETY FOR TESTING AND MATERIALS), SUBCOMMITTEE E35.21. 1978. Proposed working document for the development of an ASTM draft standard on standard practice for a laboratory testing scheme to evaluate hazard to non-target aquatic organisms. Pages 202–237 in J. Cairns, Jr., K. L. Dickson, and A. W. Maki, eds. Estimating the hazard of chemical substances to aquatic life. ASTM STP 657. Am. Soc. Test. Mater., Philadelphia.

BENTLEY, R. E., J. W. DEAN, S. J. ELLS, G. A. LEBLANC, S. SAUTER, AND B. H. SLEIGHT, III. 1978. Laboratory evaluation of the toxicity of nitroglycerine to aquatic organisms. E.G.&G., Bionomics, Wareham, Mass. 81 pp.

BIESINGER, K. E. 1975 Tentative procedure for *Daphnia magna* chronic tests in a flowing system. Fed. Reg. 40:26902–26903.

BRANSON, D. R., G. E. BLAU, H. C. ALEXANDER, AND W. B. NEELY. 1975 Bioconcentration of 2,2',4,4'-tetrachlorobiphenyl in rainbow trout as measured by an accelerated test. Trans. Am. Fish. Soc. 104:785–792.

BRODERIUS, S. J., AND L. L. SMITH, JR. 1977. Relationship between pH and acute toxicity of free cyanide and dissolved sulfide forms to the fathead minnow. Pages 88–117 in R. A. Tubb, ed. Recent advances in fish toxicology. U.S. Environ. Prot. Agency EPA-600/3-77-085. Environ. Res. Lab.-Corvallis, Corvallis, Oreg.

BROWN, S. L. 1977. Setting priorities for environmental R&D on Army chemicals. CRESS Rep. 13. SRI Int., Menlo Park, Calif. 252 pp.

CAIRNS, J., JR., AND K. L. DICKSON. 1978. Field and laboratory protocols for evaluating the effects of chemical substances on aquatic life. J. Test. Eval. 6(2):81–90.

DUTHIE, J. R. 1977. The importance of sequential assessment in test programs for estimating hazard to aquatic life. Pages 17–35 in F. L. Mayer and

J. L. Hamelink, eds. Aquatic toxicology and hazard evaluation. ASTM STP 634. Am. Soc. Test. Mater., Philadelphia.

FOOD AND DRUG ADMINISTRATION. 1959. Appraisal of the safety of chemicals in foods, drugs, and cosmetics. Assoc. Food Drug Off. U. S., Tex. State Dep. Health, Austin.

HOWARD, P. H. 1977. Prioritized guidelines for environmental fate testing of one halogenated hydrocarbon: chlorobenzene. U.S. Environ. Prot. Agency EPA-560/5-77-001. 23 pp.

JORLING, T. C. 1978. Water quality criteria. Fed. Regist. 43(97):21506–21518.

KIMERLE, R. A., W. E. GLEDHILL, AND G. J. LEVINSKAS. 1978. Environmental safety assessment of new materials. Pages 132–146 in J. Cairns, Jr., K. L. Dickson, and A. W. Maki, eds. Estimating the hazard of chemical substances to aquatic life. ASTM STP 657. Am. Soc. Test. Mater., Philadelphia.

LEE, C. C., J. V. DILLEY, J. R. HODGSON, D. O. HELTON, W. J. WIEGAND, D. N. ROBERTS, B. S. ANDERSON, L. M. HALFPAP, L. D. KURTZ, AND N. WEST. 1975. Mammalian toxicity of munitions compounds. Phase I: acute oral toxicity, primary skin irritation, dermal sensitization, and disposition and metabolism. Midwest Res. Inst., Kansas City, Mo. 103 pp.

LEE, C. C., H. V. ELLIS, J. J. KOWALSKI, J. R. HODGSON, S. W. HWANG, R. D. SHORT, J. C. BHANDARI, J. L. SANGER, T. W. REDDIG, J. L. MINOR, AND D. O. HELTON. 1977. Mammalian toxicity of munitons compounds. Phase II: effects of multiple doses. Part 1: trinitroglycerin. Midwest Res. Inst., Kansas City, Mo. 152 pp.

MACEK, K. L., K. S. BUXTON, S. K. DERR, J. W. DEAN, AND S. SAUTER. 1976a. Chronic toxicity of lindane to selected aquatic invertebrates and fishes. U.S. Environ. Prot. Agency EPA-600/3-76-046. 50 pp.

MACEK, K. J., K. S. BUXTON, S. SAUTER, S. GNILKA, AND J. W. DEAN. 1976b. Chronic toxicity of a trazine to selected aquatic invertebrates and fishes. U.S. Environ. Prot. Agency EPA-600/3-76-047.

MACEK, K. J., AND B. H. SLEIGHT, III. 1977. Utility of toxicity tests with embryos and fry of fish in evaluating hazards associated with the chronic toxicity of chemicals to fishes. Pages 137–146 in F. L. Mayer and J. L. Hamelink, eds. Aquatic toxicology and hazard evalution. ASTM STP 634. Am. Soc. Test. Mater., Philadelphia.

MCCANN, J., E. CHOI, E. YAMASAKI, AND N. B. AMES. 1975. Detection of carcinogens as mutagens in the *Salmonella*/microsome test: assay of 300 chemicals. Proc. Nat. Acad. Sci. U.S.A. 72:5135–5139.

MCKIM, J. M. 1977. Evaluation of tests with early life stages of fish for predicting long-term toxicity. J. Fish. Res. Board Can. 34:1148–1154.

MOUNT, D. I. 1977. An assessment of application factors in aquatic toxicology. Pages 183–190 in R. A. Tubb, ed. Recent advances in fish toxicology. U.S. Environ. Prot. Agency EPA-600/3-77-085. Environ. Res. Lab.-Corvallis, Corvallis, Oreg.

MOUNT, D. I., AND C. E. STEPHAN. 1967. A method for establishing acceptable toxicant limits for fish—malathion and the butoxyethanol ester of 2,4-D. Trans. Am. Fish. Soc. 96:185–193.

NATIONAL ACADEMY OF SCIENCES. 1977. Principles and procedures for evaluating the toxicity of household substances. NAS Publ. 1138. Nat. Acad. Sci. U.S.A., Washington, D.C. 130 pp.

NATIONAL RESEARCH COUNCIL. 1974. Guide for the care and use of laboratory animals. DHEW Publ. (NIH) 74–23. Nat. Inst. Health, Bethesda, Md. 56 pp.

NEELY, W. B., D. R. BRANSON, AND G. E. BLAU. 1974. Partition coefficient to measure bioconcentration potential of organic chemicals in fish. Environ. Sci. Technol. 8:1113.

PEARSON, J. G., J. P. GLENNON, J. J. BARKLEY, AND J. W. HIGHFILL. 1979. An approach to the toxicological evaluation of a complex industrial wastewater. In L. L. Marking and R. A. Kimerle, eds. Aquatic toxicology. ASTM STP 667. Am. Soc. Test. Mater., Philadelphia.

SANOCKI, S. L., P. B. SIMON, R. L. WEITZEL, D. E. JERGER, AND J. F. SCHENK. 1976. Aquatic field surveys at Iowa, Radford, and Joliet Army Ammunition Plants, vol. I: Iowa Army Ammunition Plant. Environ. Control Technol. Corp., Ann Arbor, Mich. 391 pp.

SMALL, M. J. 1977. The hazard ranking and allocation methodology: assembly of the hazard estimation data base for air and surface water pollutants. Tech. Rep. 7713, U.S. Army Med. Bioeng. Res. Dev. Lab., Frederick, Md. 43 pp. (AD A047826).

SMALL, M. J. 1978. The hazard ranking and allocation methodology: evaluation of TNT wastewaters for continuing research efforts. Tech. Rep. 7808, U.S. Army Med. Bioeng. Res. Dev. Lab., Frederick, Md. 72 pp.

SMITH, J. H., W. R. MABEY, N. BOHONOS, B. R. HOLT, S. S. LEE, T.-W CHOU, D. C. BOMBERGER, AND T. MILL. 1977. Environmental pathways of selected chemicals in freshwater systems, part I. Background and experimental procedures. U.S. Environ. Prot. Agency EPA-600/7-77-113. 81 pp.

SONTAG, J. M., N. P. PAGE, AND U. SAFFIOTTI. 1976. Guidelines for carcinogen bioassay in small rodents. NCI-CG-TR-1, Nat. Cancer Inst., Bethesda, Md. 65 pp.

SPANGGORD, R. J., B. W. GIBSON, R. G. KECK, AND G. W. NEWELL. 1978. Mammalian toxicological evaluation of TNT wastewaters, vol. I. Chemistry studies. SRI Int., Menlo Park, Calif. 317 pp.

STEPHAN, C. E. 1975. Methods for acute toxicity tests with fish, macroinvertebrates, and amphibians. U.S. Environ. Prot. Agency EPA-660/3-75-009. 61 pp.

STILWELL, J. M., D. C. COOPER, M. A. EISCHEN, M. C. MATTHEWS, AND T. B. STANFORD. 1976. Aquatic life field studies at Joliet Army Ammunition Plant. Battelle Columbus Lab., Columbus, Ohio. 169 pp.

SULLIVAN, J. H., H. D. PUTNAM, M. A. KEIRN, D. T. SWIFT, AND B. C. PRUITT, JR. 1977. Aquatic field surveys at Volunteer Army Ammunition Plant, Chattanooga, Tennessee. Water Air Res., Inc., Gainesville, Fla. 316 pp.

SULLIVAN, J. H., H. D. PUTNAM, M. A. KEIRN, D. T. SWIFT, AND B. C. PRUITT, JR. 1978. Winter field surveys at Volunteer Army Ammunition Plant, Chattanooga, Tennessee. Water Air Res., Inc., Gainesville, Fla. 168 pp.

USEPA (UNITED STATES ENVIRONMENTAL PROTECTION AGENCY). 1971. Algal assay procedure: bottle test. U.S. Environ. Prot. Agency, Environ. Res. Lab.-Corvallis, Corvallis, Oreg. 82 pp.

USEPA (UNITED STATES ENVIRONMENTAL PROTECTION AGENCY). 1972a. Recommended procedure for brook trout, *Salvelinus fontinalis* (Mitchell), partial chronic. U.S. Environ. Prot. Agency, Environ. Res. Lab.-Duluth, Duluth, Minn.

USEPA (UNITED STATES ENVIRONMENTAL PROTECTION AGENCY). 1972b. Recommended bioassay procedure for fathead minnow, *Pimephales promelas* Rafinesque, chronic tests. U.S. Environ. Prot. Agency, Environ. Res. Lab.-Duluth, Duluth, Minn.

WORLD HEALTH ORGANIZATION. 1969. Principles for the testing and evaluation of drugs for carcinogenicity. Tech. Rep. Ser. 426, World Health Organ., Geneva. 26 pp.

WORLD HEALTH ORGANIZATION. 1974. Toxicological evaluation of certain food additives with a review of general principles and of specifications. Tech. Rep. Ser. 539. World Health Organ., Geneva.

Japanese Law on New Chemicals and the Methods to Test the Biodegradability and Bioaccumulation of Chemical Substances

K. Fujiwara

Department of Environmental Medicine
Institute of Community Medicine
The University of Tsukuba, Japan

Abstract

A new law concerning the examination and regulation of manufacture, etc., of chemical substances was formulated in 1978 in Japan. The provisions of the law were divided into three parts: (1) designation of "the specified chemical substance" prior to its commercial production, import, or utilization; (2) restriction of the production, import, and utilization of the specified chemical substances; (3) listing of the existing chemical substances and screening potentially dangerous ones by government.

The criteria of the designation are as follows: (1) difficulty of natural biodegradation; (2) high concentration in bioaccumulation tests; (3) chronic toxicity in experimental animals. "The specified chemical substance" corresponds to all items from (1) to (3). If a chemical substance's biodegradation products meet the above criteria, they should be designated "specified chemical substances," too.

To test the biodegradability of chemical substances, the degree of degradation is determined from the oxygen consumption and the result of direct analysis. For the accumulation test, carp are used. For the toxicity tests, general, special, and pharmacological examinations are included.

After the unfortunate outbreak of polychlorinated biphenyls (PCB) poisoning, The Chemical Substances Control Law was formulated in 1973 in Japan. The intent of the law is to prevent hazards to the environment and human health caused by chemical substances (Chemical Product Safety Division 1977).

The provisions of the law contain the following three parts:

(1) To designate "the specified chemical substance," based upon the tests of new chemical substance prior to its commercial production, import, or utilization;
(2) To restrict the production, import and utilization of the specified chemical substance in limited areas, where "closed-cycle" of such substance from production to waste can be established;
(3) To compile a list of the existing chemical substances and to screen such substances by government.

The criteria of the designation are as follows:

(1) Whether or not a substance tends easily both to be chemically changed by microorganisms and other natural effects (biodegradability) and to be accumulated in animal body (bioaccumulation);
(2) Whether or not a substance may or might be harmful to human health through its continuous ingestion (chronic toxicity).

Those criteria were based upon the experiences of "PCB" troubles in Japan.

Test methods for these examinations will be presented in detail and data obtained will be shown in this paper.

There are more than tens of thousand chemical substances being used in various fields. The Japanese government has initiated an examination of about 800 chemicals, e.g., the alternate chemicals to PCB and others having chemical structure similar to PCB.

Discussion

The Japanese Chemical Substances Control Law is designed to prevent pollution of the environment by chemical substances which have persistence or other such properties and which may possibly be harmful to human health. It has as its purposes the establishment of a system of screening to determine, before the manufacture or import of new chemical substances, whether such substances have these properties or not, and the implementation of such regulations as may be necessary concerning the manufacture, import, and use of chemical substances having these properties (Fujiwara 1975).

Outline of the Legislation

The first part of the law deals with the regulation of new chemical substances or the preexamination system. In this system, anyone who intends to manufacture or import a new chemical substance must submit, prior to the commencement of business, a notification to the Minister of Health Welfare and to the Minister of International Trade & Industry to let the substance be subjected to the pre-examination, in which the government makes a decision whether the new chemical substance has hazardous properties or is safe.

In this case, the decision is made based in the already available informations or from biodegradability, bioaccumulation or toxicity test results from the chemical substances notification (Figure 1). If the notified substance is determined to be safe, it can be freely manufactured or imported from then on.

The second part of the law deals with the control of specified chemical substances.

A chemical substance found to have hazardous properties (persistence, bioaccumulation and toxicity) is prescribed as a Specified Chemical Substance by the Cabinet Order, and anyone who wishes to engage in the business of manufacturing or importing a specified chemical substance must obtain authorization of the government.

As for the use, it is limited to the uses specified by the Cabinet Order for each specified chemical substance respectively. Both in manufacturing and using a specified chemical substance, it is requested that a completely closed

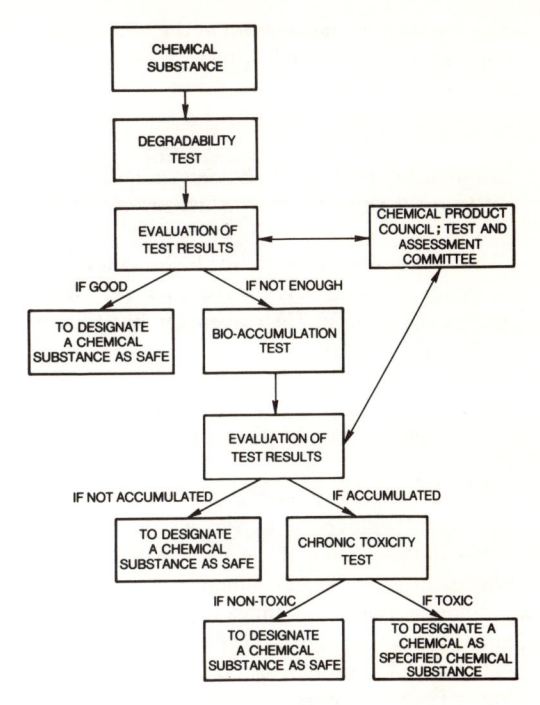

FIGURE 1.—*Procedure to designate a specified chemical substance.*

system should be adopted to avoid environmental pollution. As for products, no one can import any product proscribed by the Cabinet Order if a specified chemical substance is used therein.

The third part of the law deals with the problem of existing chemical substances. In this law, there is no provision for regulation or control of existing chemical substances but, needless to say, they should also be adequately examined with respect to their safety to man and the environment, because there may be ones with hazardous or potentially hazardous properties among them. For the time being, the government has the responsibility for examining the safety of the existing chemical substances. If an existing chemical substance is found to have hazardous properties, it will be proscribed as a specified chemical substance and proper measures for its control will be taken.

In the case that the annual amount of a new chemical substance to be produced or imported by a person does not exceed 100 kg, the Ministry of International Trade and Industry (MITI) and the Ministry of Health and Welfare (MHW) could grant the verification for such substance without the examination procedures. However, when the total annual amount of the above mentioned substance exceeds 1,000 kg, both Ministers must not grant the verification.

Method for Testing the Biodegradability of Chemical Substances by Microorganisms

The terminology used in this test method is in accordance with that used in the Japanese Industrial Standards (JIS). The reagents to be used are extra-pure or better.

Preparation of Active Sludge

Sludge sampling is done, in principle, at not less than 10 places over the whole country, chiefly in the areas where a variety of chemical substances may be consumed and discarded. Sampling is done four times a year, in March, June, September, and December.

City samples should contain 1 liter of return sludge from a sewage disposal plant. Samples from rivers, lakes, marshes, or the sea should contain 1 liter of surface water and 1 liter of surface soil from the beach which is in contact with the atmosphere.

The samples collected from the sampling sites are mixed in one container by stirring, and the mixture is allowed to stand. Floating foreign matter is removed and the supernatant is filtered through No. 2 filter paper. The filtrate is adjusted to pH 7.0 ± 1.0 with sodium hydroxide or phosphoric acid, transferred to a culture tank, and aerated.

After ceasing the aeration of this solution for about 30 minutes, about one-third of the whole supernatant volume is removed and replaced by an equal volume of 0.1% synthetic sewage (1 g glucose, 1 g peptone, and 1 g monopotassium phosphate in 1 liter of water, adjusted to pH 7.0 ± 1.0 with sodium hydroxide). This replacement process is carried out daily, and culturing is done at 25 ± 2°C.

For control of the culturing step, the following items are checked and necessary adjustments are made:

(1) The supernatant of active sludge should be clear.
(2) The active sludge, being in large flocs, must have high precipitability.
(3) If growth of active flocs is not observed, either the volume of 0.1% synthetic sewage should be added or the synthetic sewage should be added more often.
(4) The pH of the supernatant is 7.0 ± 1.0.
(5) The temperature is 25 ± 2°C.
(6) In replacing the supernatant with synthetic sewage, the solution in the culture tank must be aerated sufficiently to maintain dissolved oxygen concentrations above 5 ppm.
(7) When the active sludge is observed at 100–400× magnification, a number of protozoa of different species together with cloudy flocs must be seen.

In order to maintain fresh and old active sludges in the same activity, the filtrate of the supernatant of an active sludge in actual use in the test is mixed with an equal volume of the filtrate of supernatant of an active sludge freshly collected. Activity of active sludge should be checked periodically at least once every three months with standard substances applying the test method provided immediately below. Especially when fresh and old active sludge samples are mixed, the old active sludge must be carefully checked.

Test Method

The test uses a closed-system, oxygen-consumption measuring apparatus (Figure 2).

To make the basal culture medium, as provided in JIS K 0102-16, 3 ml each of solutions A, B, C, and D are added to water and made up to 1,000 ml. The stock solutions are created as follows:

Solution A—21.75 g dipotassium hydrogen phosphate, 8.5 g potassium acid phosphate, 44.6 g dibasic sodium phosphatedodecahydrate, and 1.7 g ammonium chloride are dissolved in water to a volume of 1,000 ml. (The pH of the solution is 7.2.)

Solution B—22.5 g magnesium sulphate heptahydrate are dissolved in water to a volume of 1,000 ml.

Solution C—27.5 g calcium chloride are dissolved in water to a volume of 1,000 ml.

Solution D—0.25 g ferric chloride hexahydrate is dissolved in water to a volume of 1,000 ml.

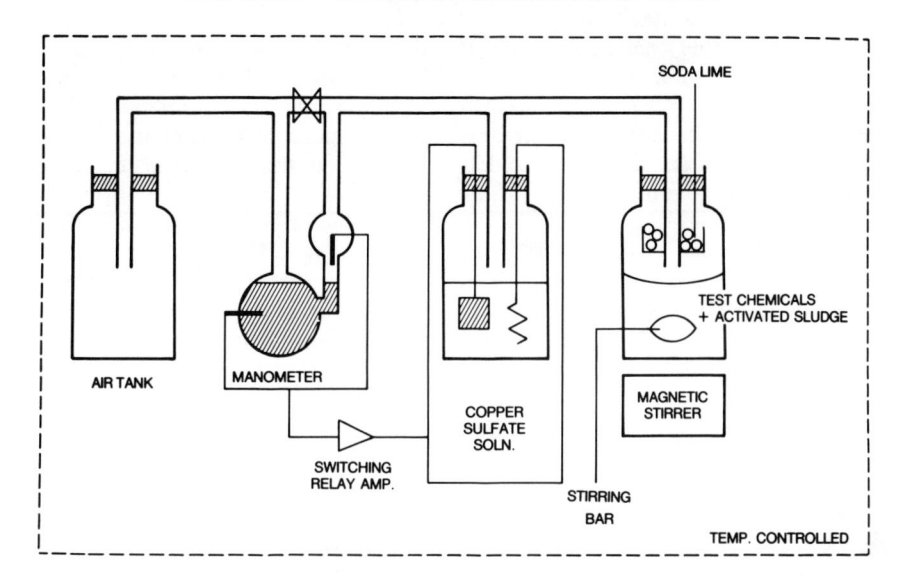

FIGURE 2.—*Closed-system oxygen-consumption measuring apparatus.*

The following testing vessels are provided, and they are adjusted to the test temperature; in the case where the test compound is not soluble in water up to a test concentration, it is pulverized as finely as possible:

(1) a testing vessel containing the basal culture medium to which is added 100 ppm (w/v) of test compound;
(2) a testing vessel for the control blank test, containing only the basal culture medium;
(3) a testing vessel containing water to which is added 100 ppm (w/v) of the test compound;
(4) a testing vessel containing basal culture medium to which is added 100 ppm (w/v) of aniline.

Inoculum is added to testing vessels (1), (2), and (4) so that the suspended matter defined in JIS K 0102-10-2 and -3 is contained in a concentration of 30 ppm (w/v). The preparation is cultured with thorough stirring at 25 ± 1°C for a given period (usually 14 days), during which changes in the amount of oxygen consumption with time are measured. After a given period of culturing is completed, the residual amount of total organic carbon also is determined.

If the percentage degradation of aniline in vessel (4), calculated from the oxygen consumption, does not exceed 40 percent, the test is regarded as invalid (Fujiwara et al. 1975).

Calculated from the oxygen consumption,

$$\text{Percentage degradation} = \frac{100(BOD-B)}{TOD};$$

BOD: biological oxygen demand (mg) of the test compound (experimental);
B: oxygen consumption (mg) of basal culture medium to which the noculum is added (experimental);
TOD: theoretical oxygen demand (mg) required when the test compound is completely oxidized.

Calculated from direct chemical assay,

$$\text{Percentage degradation} = \frac{100(Sb-Sa)}{Sb};$$

Sa: Residual amount of the test compound (mg) after completion of the biodegradability test (experimental);
Sb: Residual amount of the test compound (mg) in the blank test (vessel 2) (experimental).

In the case where a total organic carbon analyzer is used: 10 ml of the tested solution is sampled from the testing vessel and centrifuged at 3,000 × *g* for five minutes. Residual amount of the total organic carbon in the supernatant is determined on a total organic carbon analyzer.

In the case where other analyzers are used: the whole content of a testing vessel is extracted with a suitable solvent for the test compound and, after proper pretreatment such concentration, residual amount of the test compound is determined on an analyzing instrument. In this case, the analysis is made as directed in the General Rules of Analysis provided in JIS (gas chromatography, absorptionometry, mass spectrometry, atomic absorption spectophotometry, etc.).

Method for Testing the Degree of Accumulation of Chemical Substances in Fish Body

As before, terminology is that of JIS.

Test Method

Prior to initiation of the test on the degree of accumulation, an acute toxicity test is made (Figure 3). Fish used in the acute toxicity test are mature orange-red killifish *(Orizias latipes)* weighing 0.15–0.5 g. Diseased fish or those with abnormal external appearance and behavior should not be used. The test water should be fresh underground water or dechlorinated tap water sufficiently aerated so that the dissolved oxygen concentration is kept at about 7 ppm. According to the method provided in JIS L 0102-55, the test compound is dissolved in the test water; if the test compound has little solubility in water, suitable solubilizers (such as ethyl alcohol, fatty acid esters of polyoxyethylene sorbitan, etc.) can be used. The 48-hour TLm (the concentration expressed in mg/liter of the test compound at which 50 percent of the test fish will die in 48 hours) is estimated, and the TLm value is used as reference in determining the test concentration of the compound to be subjected to the test on the degree of accumulation.

The bioaccumulation test is carried out in flowing water. The apparatus is outlined in Figure 4. The aquarium should be glass, clean, and of sufficient volume for rearing the test fish. Other apparatus used for water supply or dilution of the test compound should be glass and as clean as possible. Use of plastics should be limited to the parts where it is inevitable, such as joints.

Carp *(Cyprinus carpio)* weighing 20–40 g and measuring about 10 cm in standard length are used as test fish. They are reared in a suitable pond, and weakened fish and those showing other abnormal signs should be eliminated. Then, external or internal pathogenic parasites are exterminated by sterilizer bathing or medication to keep the fish in good body condition, and after that they are transferred to an acclimation tank.

If the test temperature is higher than the water temperature of the fish pond, fish are acclimated for not less than a day in a temperature higher by 5°C or less than the water temperature of the fish pond. Then the temperature is gradually raised at the rate not more than 3°C/day and finally the fish are reared in the test temperature for 5 to 7 days.

If the test temperature is lower than the water temperature of the fish pond, fish are acclimated for not less than a day in a temperature lower by 3°C or less than the water temperature of the fish pond. Then the temperature is gradually lowered at the rate not more than 2°C/day and finally fish are reared in the test temperature for 7 to 10 days.

In the meantime, the fish with damage to the gills or the skin or that are diseased are eliminated. It is desirable that the water in the fish pond and acclimation tank is always flowing. The carp in the acclimation tank in which more than 5 percent of all fish have weakened or died within one week before the start of the test should not be used.

Fish are fed pelleted feed 2 to 3 times daily. Compounded feed contaminated by residual agricultural chemicals or other chemicals that might cause serious effects on the test must not

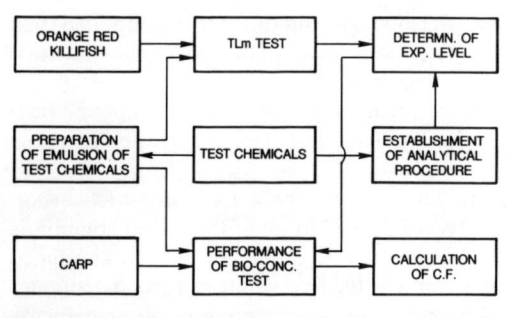

FIGURE 3.—*Block diagram of the biocencentration test.*

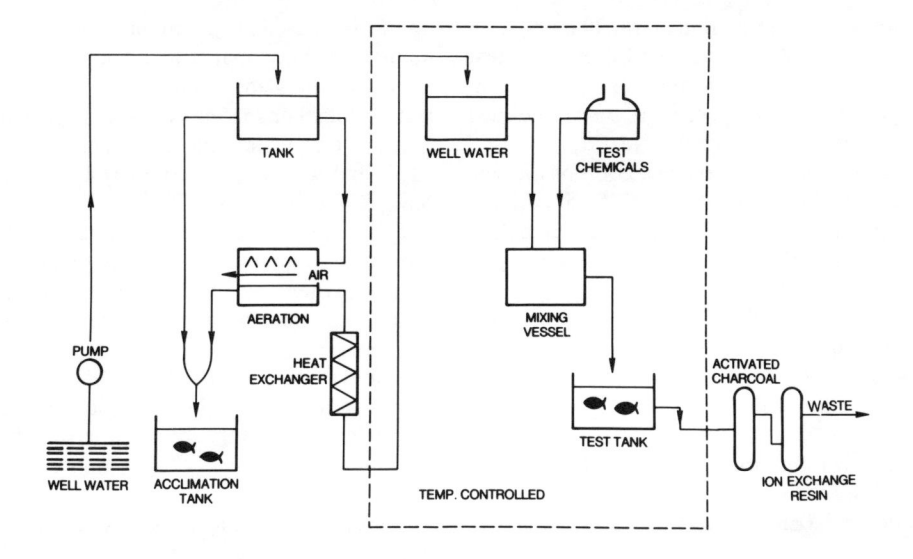

FIGURE 4.—*Aquatron in the Chemical Biotesting Center.*

be used. In order to avoid any leftover, the feed should be given slowly so that the portion of the feed floating on the water surface is entirely eaten by the test fish before it reaches the bottom of the aquarium.

The test water is as for the acute test with killifish, above. Two concentrations of the test compound, one of them as low as possible within the analytical limit, the other 10 times higher, should be prepared. Levels of 1/100, 1/1,000 and 1/10,000 of the 48-hour TLm for killifish will serve as a guide. As before, if the test compound has little solubility in water, suitable solubilizers are used.

The test is carried out in each of the two concentration levels, and at the same time blank test is made in the water in which the test compound is not dissolved. During the test period, test water is controlled so that the concentration of dissolved oxygen is kept at about 7 ppm. Water temperature is kept at 25 ± 2°C.

Healthy test fish are reared, under the conditions outlined above, in the aquatron (Figure 4). Excretions of the carp and sludges on the wall of the aquarium should be removed about once a day throughout the test period. The test should be continued for 8 weeks, in principle. However, in the case that the concentration factor calculated by the equation below exceeds a given value, or in the case it gives a maximum value for either of the two levels of test concentration, the test may be discontinued.

Every one or two weeks, two to three test fish are taken out not less than ten hours after feeding and subject to analysis. The fish from the water concentration level in which the test fish have showed distinct abnormal signs during the test should not be used for analysis.

The surface of fish body is blotted dry with gauze and each fish is weighed accurately to 0.1 g. Next, the whole body of each fish, in principle, is homogenized and a suitable pretreatment such as extraction and concentration, depending upon the chemical properties of the test compound, is done to prepare a sample for quantitative analysis. The analysis is made as directed in the General Rules of Analysis provided in JIS (gas chromatography, absorptiometry, mass spectrometry, atomic absorption spectrophotometry, etc.).

The degree of accumulation, expressed by concentration factor, is calculated by the following equation:

$$CFn = \frac{Fn - FB}{W} ;$$

CFn: concentration factor after n weeks;
 Fn: concentration of the test compound in the fish body after n weeks in the test period;
 FB: arithmetic mean concentration of the test compound in the fish body at the start of,

and at the termination of, the blank test; in the case CFn is calculated during the test period, the concentration of the test compound in fish bodies before the start of test may be quoted as tentative FB;

W: arithmetic mean concentration of the test compound in the aquarium, from periodic measurements; however, the calculated concentration may be used instead if necessary.

Finally, draw a correlation curve of CFn values against time.

General and Special Toxicity Tests, Pharmacokinetics Study, and General Biological Tests of Chemical Substances

General Toxicity Tests

Acute toxicity tests — All the animals employed in the experiment are checked for acute toxicity symptoms for not less than a week to determine the lethal dose. Not less than two species of animals, such as mice and rats, both males and females, are used. The test compound is administered to animals orally and parenterally.

Chronic toxicity tests — All animals employed in the experiment are checked for their functional and morphological changes; the maximum non-influential dose, the minimum toxic dose, and the definitively toxic dose are determined. Not less than two species of animals, such as mice and rats, both males and females, are used. Administration of the test compound is started immediately after weaning of the experimental animals and administration is continued, in principle, for the major part of their lives.

Special Toxicity Tests

Test on the effect on reproduction and subsequent generations — All the experimental animals that conceive are allowed to give birth to newborn animals on which observations are made to estimate the maximum non-influential dose. One or more species of experimental animals, such as mice and rats, and not less than two generations as the subsequent generations, are used. Oral administration of the test compound is started, three months or more before mating, to both males and females of experimental animals and is continued throughout the test period. The mating is made, in principle, by housing one male and one female animal in a cage, and it should be confirmed whether they have copulated or not.

Teratogenicity test — The prepartum dams of the animals employed in the experiments, and all fetuses in them, are checked for teratogenicity to estimate the maximum non-influential dose. Not less than two species of animals, such as mice, rats, and rabbits, are used. The test compound is administered orally, in principle, at the stage when the organs of the fetuses are being formed.

Tumorigenicity test — All animals employed in the experiments are checked for the occurrence of tumors. One or more species of experimental animals, such as mice and rats, both males and females, are used. The maximum tolerable doses are administered orally throughout their lives.

Pharmacokinetics Study

Animals employed in the experiments are checked for absorption, distribution, accumulation, metabolism, and excretion of the test compound. Not less than two species of experimental animals, such as rats, rabbits, and dogs, both males and females, are used. The test compound is administered orally.

General Biological Tests

Pharmacological effects of the test compound on nervous systems, respiratory organs, and circulatory systems are estimated. Cultured cells and cultured tissues are also to be checked for their functional and morphological changes due to the test compound. Experimental animals such as mice, rats, and rabbits are used. The test compound is administered orally or by other routes.

Examination Results

Results obtained by the procedure of this examination through four years are shown in Table 1. Among 559 new chemical substances, 216 were designated as nonhazardous. After the

TABLE 1.—*Results of examination of new and existing chemical substances in Japan.*

Procedure or decision	Fiscal year				
	1974	1975	1976	1977	Total
New substances					
Notified	241	74	104	140	559
Production	128	43	73	81	325
Import	113	31	31	59	234
Designated as safe	16	37	65	98	216
Production	8	22	51	58	139
Import	8	15	14	40	77
Specified chemical substance	1				1
Existing substances					
Biodegradation test	55	101	97	71	324
Designated as safe	18	37	25	15	95
Bioaccumulation test	15	16	31	47	109
Designated as safe	6	13	24	43	86
Total designated as safe	24	50	49	58	181

tests of existing chemical substances, 181 substances were designated as safe among 324 substances examined.

Conclusions

From our experiences through four years, test items and test systems applied in Japan are deemed to be appropriate generally. In regard to the test methods and evaluation of test results, however, we need to make efforts to improve and to develop them so as to possibly simulate the real behavior of chemicals in natural conditions. One of these efforts we are discussing is an anaerobic biodegradation test, and a procedure and apparatus for such purpose is now being devised.

The volume, the distribution, and the treatment, as well as the properties of a substance, are important factors in assessing an environmentally hazardous agent. Therefore, it is necessary to allow some flexible control, taking into account the volume, usage, and treatment of a chemical substance. Additionally, in accordance with the progress of civilization, we need to consider how to improve our legal criteria in the future.

Acknowledgment

We should like to express our thanks to Mr. Kawasaki, Director, Chemical Products Safety Division, MITI, Japan, for information about the internal situations concerning safety problems with chemical substances.

References

CHEMICAL PRODUCTS SAFETY DIVISION, MINISTRY OF INTERNATIONAL TRADE AND INDUSTRY. 1977. The chemical substances control law in Japan. Basic Ind. Bur., Minist. Int. Trade Ind., Japan.

FUJIWARA, K. 1975. Studies on the chemical substances control law. J. Org. Synth. Chem. 33:5.

FUJIWARA, K., T. SUGIYAMS, AND K. KOIKE. 1975. Biodegradation of linear alkylatesulfonates in river model. Jap. J. Hyg. 29:552–557.

The Use of the Concentration-Response Relationship in Assessing Acute Fish Toxicity Data

R. LLOYD

*Ministry of Agriculture, Fisheries and Food
10, Whitehall Place
London, England SW1A 2HH*

Abstract

The use of the concentration-response relationships obtainable in fish or invertebrate acute toxicity tests can provide valuable information in assessing the hazard of a chemical. Critical test requirements in generating the concentration-response relationship include consistency of water quality and the concentration of the test substance during the test. Continuous-flow assays generally meet these requirements. By analyzing the shape of log plots of LT50's, information about the further test requirements of a sequential hazard assessment scheme can often be obtained. For example, shallow curves may indicate that a chemical is likely to accumulate in the test organism's tissues, suggesting the need for bioconcentration testing.

The use of application factors which can be applied to 96-hour LC50 data from standard acute toxicity tests is governed by the relationship between the sensitivity of the species being tested and the most sensitive species to be protected; effects changes in water quality characteristics have on the toxicity of the chemical; the relationships between the minimum 96-hour LC50 and the concentration likely to be harmful to the most sensitive stage in the life cycle; and toxic form and stability of the chemical.

For any new chemical, the primary information obtained within an aquatic hazard evaluation scheme is from a short-term acute toxicity test; the species of test animal is usually a fish, but increasingly an invertebrate (commonly *Daphnia* sp.) is included. Data from such short-term tests with fish are usually expressed as LC50's for exposure periods of between 24 and 96 hours. Occasionally further statistical analyses of dubious utility are carried out on the data to provide confidence limits for those concentrations and the slope function. However, it has been recognized for some years that the shape of the concentration-response curve can be obtained with little extra effort from these tests and that this information can be of considerable value (Sprague 1973; Lloyd 1977). With the increasing demand for ecotoxicity testing of new chemicals, there is a need to extract the maximum amount of information from each test in order to conserve time and resources.

It is possible that the reason why this extra information is not repeated more frequently is that the toxicity tests would then have to be carried out with greater accuracy; the critical test requirements are that the water quality, and the concentration of the test substance, should be held as near constant as possible during the test period. In practice, this requires the use of a continuous-flow type of test apparatus, since reductions in the concentration of test substance, or in the dissolved oxygen content of the water, may give spurious inflections to the concentration-response curve during the test period.

Concentration-response curves can be constructed from a series of fixed-period LC50's (3, 6, 24, 48, and 96 hours) or from more frequent observation, in which case it is possible to calculate LT50's for each concentration in which more than 50% of fish die. It is of interest to note that the standard fish toxicity test protocol which is being formulated by the International Standards Organization (1978) proposes that the report of the test results should include both the 96-hour LC50 and the concentration-response curve.

Concentration-Response Curves

It is customary to plot log exposure concentration against log median response time; in this way, the relationship obtained from acute toxicity test data can range from linear, as curve C in Figure 1, to strongly curvilinear as in curve A. Occasionally, more complex double curves can be obtained. Because of the valuable information provided by the concentration-response relationship, the validity of a curve obtained has to be proved. False inflection of the curve may be caused by a decrease in concentration of pollutant, or by volatility of the test substance, or by its degradation or absorption onto test container surfaces. Therefore, confidence in a type of curvilinear response depends on evidence that the concentration of pollutant has remained constant (or near constant) during the test period. Verification of the shape of simple relationships can sometimes be obtained by inspection of the standard deviations of individual survival times from the mean for each LT50, which will increase with an increasing slope of the curve, or of the standard deviations of concentration from each fixed-period LC50, which will decrease with increasing slope of the curve.

Use of the Concentration-Response Curve Within a Sequential Hazard Assessment Scheme

It is assumed that one of the first steps in a sequential hazard assessment scheme will be to carry out an acute toxicity test with a chosen species of fish. From the data obtained, a reasonable (or not unreasonable) over-prediction of the minimum concentration likely to be harmful to the most sensitive species of fish (or the most sensitive of a particular group of fish species) has to be estimated. There is no generally accepted application factor which can be applied to a 96-hour LC50 from a standard acute fish toxicity test to estimate the minimum concentration of a chemical which is harmful to fish. It has been suggested (EIFAC 1975) that concentrations less than four orders of magnitude lower than the acute toxic concentration (96-hour LC50 $\times$ 10^{-4}) are likely to be harmless to fish populations. Experience shows that this approach would considerably over-estimate the potential harmfulness of many chemicals, especially those which have a concentration-response relationship similar to that of curve A in Figure 1.

It seems appropriate, therefore, that the application factor should be examined in greater detail, and for this purpose, it has been assigned to four components as follows:

(1) The relationship between the sensitivity of the species under test and that of the most sensitive species to be protected. The closer the species similarity, the smaller the factor.
(2) The effect of changes in water quality characteristics on the toxicity of the chemical or on the sensitivity of fish to the chemical; for example, water hardness, pH, salinity, and temperature.
(3) The relationship between the minimum 96-hour LC50 derived from (1) and (2), above, and the minimum concentration likely to be harmful to the most sensitive stage in the life cycle.
(4) The toxic form and stability of the chemical.

The main effect of the first two of these variables is to move the position to the concentration-response curve in relation to the axes. Ideally, the effects of these variables should be measured in terms of changes in the threshold LC50, since the use of shorter exposure times with greater LC50's can give misleading comparative data. However, it is possible that factors to be applied to the 96-hour LC50 can be obtained by using an approach basically similar to that adopted by the Environmental

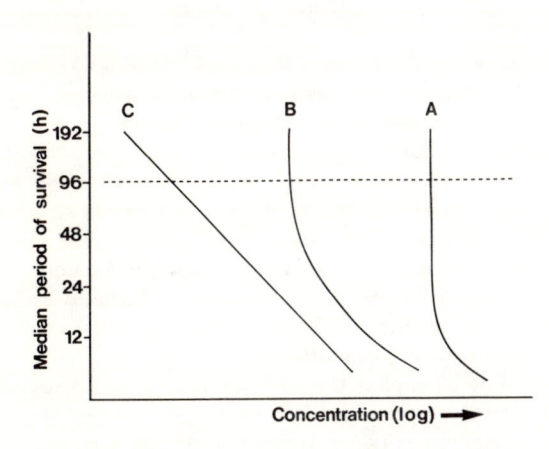

FIGURE 1.—*Three types of concentration-response curves obtained in acute fish toxicity tests (from Lloyd 1977)*

Protection Agency as outlined in the Federal Register (Jorling 1978), in which a wide range of comparative data was analyzed. Such an approach may be valid in this context, since the aim is to provide an approximate over-prediction (i.e., with a margin of safety) as a first stage in an assessment programme, in contrast to the setting of definitive water quality standards. It should be pointed out, however, that species of fish shown to be relatively resistant in short-term tests may be more sensitive in the long term. Nevertheless, experience suggests that as a first approximation, a factor of 5 may be appropriate to allow for interspecies sensitivity where an indigenous species has been used for the test, and a factor of 10 for a nonindigenous species (for example, zebrafish *Brachydanio rerio*). Evidence of the effect of water quality characteristics may be obtained from citing toxicity tests carried out in waters of different hardness or salinity; in the absence of such data, a factor of 10 could be used.

Factors appropriate to component (3), above (i.e., the relationship between the acute LC50 and the minimum concentration likely to be harmful to the most sensitive stage in the life cycle), will depend in part on the shape of the concentration-response curve. For example, curve A becomes parallel with the time axis after a short exposure period and a threshold LC50 can be firmly established. Such a response indicates that the test species of fish can detoxify the chemical or adapt to it at low concentrations with reasonable facility, and here an application factor of 10 might not be unreasonable. A concentration-response relationship similar to that shown as curve B indicates that the chemical is not so readily detoxified or adapted to, and that the 96-hour LC50 is close to the threshold concentration. This is probably the most common type of concentration-response curve with fish, and in this case an application factor of 20 might be appropriate. For curves intermediate in shape between B and C, a factor of 100 might be reasonable if it can be assumed that the threshold LC50 is within an order of magnitude of the 96-hour LC50.

No single application factor can be applied to a linear relationship as in C, although it may be reasonable to extrapolate the line to estimate the LC50 for the maximum life-span of the test fish, and to apply an application factor of 10 to that concentration (Lloyd 1978).

It is assumed that, for predictive purposes, the shape of the concentration-response relationship found for a test species of fish with a specific chemical will be common for other fish species also, even though the position of the curve with respect to the time and concentration axes will be changed. There is some validity for this general assumption in that there are very few examples of significant interspecific differences in concentration-response curves obtained for a chemical.

Following this line of reasoning, the minimum concentration of a chemical likely to be harmful to fish can be expressed as:

$$\frac{\text{96-hour LC50}}{\text{Factors for components (3)} \times (2) \times (1)}.$$

The factors assigned to each component will depend on a critical examination of all pertinent existing data, judgment, and experience; no rigid guidelines can be set down at this time.

The modifying effects on toxicity of the chemical and physical properties of a chemical (4) other than those included in (2), above, are not reflected in the concentration-response relationship, and are therefore outside the scope of this paper. In general, however, these properties are more likely to reduce than to increase the toxicity of the chemical in natural aquatic ecosystems (although adsorption onto suspended matter may transfer the toxicity from the aquatic to the benthic phase), and if necessary, these factors can be taken into account in the subsequent stages of the assessment procedure.

Further Information Obtainable from Concentration-Response Curves

The shape of the concentration-response curve within the period of acute testing can give information on the further test requirements if the preliminary assessment of the minimum concentration of chemical likely to be harmful to fish is lower than the maximum predicted environmental concentration.

For example, it is obvious that the first requirement for chemicals which give curves intermediate between B and C in Figure 1 is that they should be tested over an extended exposure period to ascertain whether a threshold LC50 exists. Shallow curves may indicate that the chemical is likely to accumulate in the tissues of fish to concentrations well in excess of the ambient value, and tests to determine bioaccumulation may be initiated at an early stage; these may complement an examination of data on the solvent-water partition coefficient for the chemical.

Conclusions

It is realized that the use of information from concentration-response curves can only be incorporated within those hazard assessment schemes that have a flexible approach to the acquisition of further data, and therefore it may not be capable of incorporation within a rigid assessment protocol. Nevertheless, the information which can be obtained is of obvious value and every attempt should be made to incorporate the acquisition and analysis of such data within sequential testing assessment programmes.

Finally, although the principles outlined in this paper have been linked with acute fish toxicity tests, they are probably applicable to acute invertebrate toxicity tests also.

References

EIFAC (EUROPEAN INLAND FISHERIES ADVISORY COMMISSION). 1975. Report on fish toxicity testing proceduces. EIFAC Tech. Pap. 24. 25 pp.

JORLING, T. C. 1978. Water Quality Criteria. Fed. Regist. 43(97):21506–21518.

INTERNATIONAL STANDARDS ORGANIZATION. 1978. Recommended changes to the ISO proposed flow-through test procedure ISO/TC 147/SC5 WG 3 Doc 18—Document presented to ISO Meeting, Ottawa, May 1978.

LLOYD, R. 1977. Are short-term fish toxicity tests a dead end? Paper presented to Section K of the British Association for the Advancement of Science, Aston University, Birmingham. 7 pp.

LLOYD, R. 1978. Toxicity tests with aquatic organisms. Paper presented to the 6th FAO/SIDA Workshop in aquatic pollution in relation to protection of living resources, Mombasa, Kenya.

SPRAGUE, J. B. 1973. The ABC's of pollutant bioassay using fish. Pages 6–30 in J. Cairns and K. L. Dickson, eds. Biological methods for the assessment of water quality. ASTM STP 528. Am. Soc. Test. Mater., Philadelphia.

Discussion Synopsis—
Hazard Assessment Approaches

J. R. Duthie, *Chairman*

Wesley Birge, N. T. de Oude, David Hansen,
Richard Lloyd, Gordon Loewengart, and James Reisa

Proliferation of hazard assessment schemes has occurred worldwide over the last several years. Several aquatic hazard assessment schemes were presented, reviewed, and compared at the 1977 Pellston Workshop on the Application of Aquatic Toxicity Testing Methods as Predictive Tools for Aquatic Hazard Evaluation (Cairns et al. 1978). Included were the American Society for Testing and Materials (ASTM) draft scheme, applicable broadly to chemicals; the American Institute of Biological Sciences (AIBS) procedure, designed for specific application to pesticides under the Federal Insecticide, Fungicide, and Rodenticide Act (FIFRA); and an industrial approach used by Monsanto. At the 1978 workshop, more than 15 schemes in various stages of development were presented or discussed, including those by Lloyd (England), Fujiwara (Japanese New Chemicals Law), Lundahl (France), Lee (Unilever, UK), USEPA Toxic Substance Control Act (TSCA), Hueck (Netherlands), and Pearson (US Army).

Many more schemes could easily be produced and it should be clear that the purpose of this session was not to generate another new plan. Rather, in 1978 we addressed whether the new plans that were considered confirm or deny the general characteristics of aquatic hazard assessment programs defined at the Pellston meeting. Further, with this updated and expanded base of assessment approaches, can we identify commonalities and differences and reasons for them; can we define the current state of the art; and can we influence future development by highlighting evolving areas?

Comparison of Schemes

The comments and questions arising in the general discussion following the presentation of new hazard schemes led the synthesis committee to establish specific parameters and an overall framework by which plans could be analyzed and compared. The parameters initially identified the intended *scope* of each plan, including its objective, expected utilization, breadth of biological systems, and media covered. Secondly, the presence or absence of major design *components* was analyzed, including use of environmental fate data, comparison of exposure concentration and effects concentration, the phased or tier approach, identification of decision points, the specificity of criteria for decisions, and the alternatives for decisions. Finally, an attempt was made to assess the *utility* of the scheme by considering the testing resources (cost) and the level of risk associated with utilization.

In Table 1, these parameters are organized into a matrix and are used to compare two reference schemes (ASTM) and AIBS/FIFRA (partial) with the five schemes presented in this Session. The presence of an "O" may indicate a complete lack of consideration for the specific factor or that it was only a minor consideration in that hazard evaluation scheme. A major objective of six of the schemes was to assess the hazard of substances in the aquatic environment. The seventh, the Japanese scheme, was designed to assess the impact of substances on man via bioaccumulation in aquatic species. However, since this system was designed primarily to assess human health effects and involves only a limited aquatic hazard assessment, it will not be a part of further comparisons.

The objective of environmental schemes was either to evaluate the environmental hazard of new or expanded use chemicals, to assess the hazard of special-use groups of chemicals (e.g., pesticides), or to develop water quality criteria.

TABLE 1.—*Comparison of hazard evaluation schemes by the American Society for Testing and Materials (ASTM), the American Institute of Biological Sciences/Federal Insecticide, Fungicide, and Rodenticide Act (AIBS/FIFRA), and present symposium authors. X denotes parameter is included in the scheme. O denotes parameter receives minor or no consideration in scheme.*

Comparison parameters	ASTM	AIBS/FIFRA	Lloyd	Lundahl	Lee	Pearson	Fujiwara
Scope							
1. *Objective*							
a. Assess new chemicals/new uses	X	O	X	X	O	O	X
b. Assess special new chemicals	X	X	X	X	X	O	X
c. Set water quality criteria	O	O	X	O	O	X	X (?)
d. Assess in-use chemicals	O	O	X	O	O	X	X
2. *Utilization*							
a. Regulation or restriction	O	X	X	X	O	X	X
b. Guidance for new product development/manufacture	X	O	O	X	X	O	O
3. *Biological systems*							
a. Human health (public or occupational)	O	X	O	O	O	X	X
b. Aquatic species	X	X	X	X	X	X	X
c. Wildlife	O	X	O	O	O	X	O
d. Terrestrial plants	O	X	O	O	X	X	X
4. *Media*							
a. Water	X	X	X	X	X	X	X
b. Air	O	X	O	O	O	X	O
c. Land	O	X	O	O	X	X	O
Design Components							
5. *Utilizes chemical/physical data for:*							
a. Fate/environmental concentration calculation	X	X	X	X	X	X	O
b. Test selection/design	X	X	X	X	X	X	O
6. *Utilizes use and disposal patterns*	X	X	X	X	X	X	X
7. *Compares environmental concentration to effects concentration*	X	X	X	X	X	X	O
8. *Identifies clear tiers/phases*							
a. Screening	X	X	X	X	X	X	O
b. Predictive	X	X	X	X	X	X	O
c. Confirmative	O	X	X	X	X	X	O
d. Monitoring	X	X	O	X	X(chemical)	X	O
9. *Identifies periodic decision points*	X	X	X	X	X	X	O
10. *Provides specific decision criteria/triggers*	X	X	X	X	X	X	X
11. *Provides decision options (go/no go/further study)*							
a. At early points	X	Study/accept	X	Reject/study	Reject/study	Reject/study	O
b. At several points	X	Study/accept	X	Reject/study	Accept/study	Reject/accept/study	O
c. At conclusion	X	Study/accept/reject/restrict	X	Accept/restrict/reject	Accept/reject/study	Estimate criterion	Accept/reject
12. *Identifies points for data transfer to/from other systems*	X	X	O	O	X	X	O
13. *Involves sophisticated toxicological principles*							
a. Sensitive species	X	X	X	O	X	X	O
b. Mode-of-action considerations	O	O	O	O	O	O	O
c. Prediction of sublethal effects	X	X	X	X	X	X	O

TABLE 1.—*Continued.*

Comparison parameters	Source of scheme						
	ASTM	AIBS/ FIFRA	Lloyd	Lundahl	Lee	Pearson	Fujiwara
d. Evaluates laboratory effects for ecological significance	O	X	O	O	O	X	O
e. Prediction of community effects	O	O	O	O	O	X	O
Utility							
14. *Time and cost requirements*	Note 1	Note 1	Depends on objective	Note 1	Note 1	High	Low
15. *Level of risk*	Note 2	Note 2	Note 2	Note 2	Note 2	Note 2	Note 2

Note 1—Effort is material-specific and dependent also on investigator's judgments (see text).
Note 2—Level of risk depends not only on kinds and amounts of data available to decrease uncertainty of the hazard estimate, but also the kind of environmental risk judged to be acceptable (see text).

The thrust of the objective appears to be a very important factor that influences differences in the schemes. The schemes also differed in content, depending on whether they were utilized to make regulatory decisions or to provide product development and manufacturing guidance.

The scope and content of the workshop was aquatic hazard assessment, thus this evaluation process is focused particularly on the biological systems tested, and on the defined points of data transfer to and from nonaquatic systems.

All those hazard assessment schemes designed to estimate the impact of chemicals on the aquatic environment utilized common information and contained common elements within their designs. All schemes utilize data on effects of chemicals on aquatic species. In addition, they all use physical-chemical data on the compound along with its actual or projected use and disposal patterns. These latter data were then used to estimate expected environmental concentrations which, when compared to the aquatic effects concentrations, formed the basis of the hazard estimate.

Review of the table shows that all the schemes to estimate environmental hazard have the following design elements in common: (1) distinct tiers or phases; (2) some decision criteria; (3) provisions for periodic review and decision points; and (4) decision options.

While all contained separated tiers or phases, their commonality generally was restricted to screening and predictive phases. The specific test requirements, even within these common phases, were variable in type and number, and depended upon the objectives of the respective schemes. Decision criteria were common to all schemes. Quantitative criteria were generally associated only with the screening phase and were highly variable. Qualitative criteria were present in all phases and became dominant during the later phases. The inability or reluctance to provide specific quantitative decision criteria in later phases is believed due to the difficulty in anticipating and integrating the larger and more complicated data set that will be available. Advance selection within the data set of the key factors considered necessary to evaluate hazard, and a predecision on their appropriate weight in making an overall final objective assessment, appears to be antipathetic to the responsible scientist. Provisions for periodic review and decision points were also common to all schemes and were generally associated with the completion of discrete tasks or phases. These periodic decision points appeared to be related to the authors' desires to identify in the shortest time possible those compounds with high potential for adverse impact. All schemes also had provisions for various decision options at intermediate milestones and at the completion, as can be seen from Table 1. The specific options are tied directly to the objectives of the respective scheme.

A comparison of the utility of schemes in the abstract sense proved impracticable. Since each scheme had a specific and usually different purpose, the effort necessary to evaluate chemicals was related to the breadth of that purpose. Further, since the properties of a chemical should affect the kind of testing, a comparative analysis of effort or costs would necessarily require application against specific chemicals. While such a trial to obtain comparative estimates might prove a useful exercise, it could be subject to considerable error in the confirmatory stages where the requirement for extended testing will depend upon the expertise and judgment of the investigator.

In the same way there are problems of evaluating relative risk to the environment arising out of use of the various plans. The uncertainty associated with a final hazard estimate and the degree or kind of environmental risk judged to be acceptable are neither separated nor quantified in any of the plans. While the uncertainty of the final hazard assessment may be a function of a specific plan, the acceptability of the risk to the environment is judgmental in nature, and normally involves social as well as technical input.

The Use of Sophisticated Toxicological Principles in Hazard Assessment Schemes

The need to apply schemes more or less broadly to groups of unknown and vastly different chemicals leads to test simplifications that cause concerns in the basic research community and also to those non-task oriented but concerned ecological awareness groups.

For the most part, those persons directly involved with test execution or regular review tend to favor use of species and methods with which they are familiar and for which there are established comparative data bases. Both experience and reference data are important in data interpretation.

A review of the present hazard evaluation schemes raised questions about the level of sophistication required in the aquatic testing methodology. Such questions arise because more in-depth procedures are almost routinely applied in mammalian toxicology and because aquatic toxicology must necessarily extrapolate from effects on a few species to effects on a complex ecosystem.

Specific suggestions were made to consider more fully species selection, mode of action studies, determination of sublethal effects, and evaluation techniques to estimate potential ecological impact and to predict community effects. Some additional details relative to these suggestions follow.

Species Selection

Ideally the selection for testing would meet the following criteria:

(1) Species is sensitive to type of toxicant to be tested.
(2) Species is representative of the food chain or trophic level.
(3) Organisms are of known genetic history and comparable to wild species.
(4) Species' nutritional and environmental requirements are known.
(5) Species is of commercial or recreational importance.

The scientific desirability of fulfilling these requirements is recognized in research or in advanced phases of testing for determining water quality criteria for selected materials. Operationally, for screening many materials, a complete adherence to all ideal factors will not occur.

Utilization of Chemical, Physical, and Biological Properties of Test Materials

Concerns have been raised that factors affecting kinetics of uptake, the mode of action, etc., may not be fully considered in the test species used or in the selection of reference chemicals for which there is real world monitoring information. Writing such special considerations into a general test scheme is of course difficult and probably undesirable.

Prediction of Sublethal Effects of Ecological Significance

Most schemes of hazard evaluation rely on single species laboratory tests and measure lethality or inhibition of growth and reproduction. Sublethal effects on health or behavior which might affect species succession, migration, immigration, and predation are usually not measured so that effects on ecosystem function, population dynamics, or community structure are not fully assessed. There are provisions in some schemes to utilize multispecies ecosystem models or small-scale field tests, or field monitoring to confirm laboratory predictions where appropriate.

Certain "effect indicators" utilizing monitored sublethal responses, animal biochemistry, and pathology are being developed and have been proposed for broader use with environmental species.

Acute tests measuring lethality, when combined with knowledge of physical and chemical environmental interaction, allow a reasonable prediction of mortality in the real world. Difficult to quantitate is the importance of a reduction in growth or fecundity or of more subtle physiological effects for their ecological significance in population dynamics or in community structure or function.

Assessment of Sophisticated Testing

Both overall hazard assessment schemes and their specified or suggested testing requirements should be sufficiently flexible to accommodate new methods or new species models, when their reliability and utility can be demonstrated. At present and in the near future, there is a considerable backlog of testing required to establish the basic toxicological and fate information for new compounds and even for many already in use. This main-line effort may appropriately need to bypass more sophisticated, speculative, or unqualified procedures which could provide "interesting" but more difficult to interpret data.

In the continuing development or modification of hazard evaluation schemes, those places in the scheme and those properties of compounds should be identified which would benefit from the application of more sophisticated toxicological protocols. The utility of new methods to provide useful information must be carefully evaluated with due regard to cost and kind of manpower available. Such evaluation means that the application of many such procedures to many chemicals will be judged unnecessary and unlikely. In lieu of the application of these more sophisticated testing procedures which might identify more subtle toxicological effects, current practice is to use appropriate margins of safety with the simpler predictive tests.

International Considerations

The passage and implementation of the Toxic Substances Control Act in the US has been accompanied by parallel initiatives in other countries, which have developed, or are in the process of developing, hazard assessment schemes in support of national legislation. The workshop was made aware of developments in Japan and in the European Economic Community (EEC) and its member states. It was apparent that these hazard assessment schemes vary in scope and complexity. For example, the design of the Japanese scheme is simple in that, initially, only three tests (which vary in complexity) are required, and a decision to continue testing is made after each test. This scheme is already operational and a large number of new chemicals have been assessed for potential public health risks arising from the consumption of contaminated fish, as well as for other properties.

Other national proposals for hazard assessment schemes for environmental protection, like those in the US, are still in the developmental stage. For example, the EEC proposals (in support of Annex VII to the Sixth Modification to the 1967 Directive on Dangerous Chemicals) include a comprehensive list of information requirements to be generated for the purposes of the assessment programs. In this way, authorities will be notified of "dangerous" chemicals and appropriate action taken in respect to packaging, labeling, and possible restrictions of use.

In this respect, the objectives of the EEC and US legislation are basically similar and it may be expected that the supportive hazard assessment programs which will be developed will also have several common aspects. However, it is recognized that there may well be different emphases in response to specific national environmental protection requirements, the several national legislative frameworks, and national popular and special interest group opinions.

These differences could lead to major duplication of effort in support of chemicals developed for international markets. Such duplication is already apparent in international pesticide registration. Extension of such duplicated data requirements to all new chemicals could lead to a heavy burden being placed on existing or projected testing facilities and resources.

There is an immediate requirement to ensure that this potential situation does not arise; several aspects of assessment schemes are amenable to international coordination. These include (a) the exchange and international availability of pertinent information, (b) the recognition of common aspects among the hazard assessment schemes, leading to the acceptance by national registration authorities of comparable data generated in other countries, and (c) an agreement on the extent of testing required before a chemical is accepted in order to avoid unnecessary over-testing.

There is general scientific acceptance that the choice of test species representing different trophic levels and different phyla will increase considerably the value of the data base generated for a chemical. This principle should be borne in mind when standardization of test species is being considered, particularly where the number of test species is limited, and over-representation of species from the same order or trophic level should be avoided.

The Organization for Economic Cooperation and Development (OECD) is already active in international coordination of hazard assessment schemes and testing requirements, and this initiative is welcomed. International harmonization in this field is urgently needed.

Summary and Conclusions

On a broad geographical basis, many hazard evaluation schemes now exist. They provide a framework for logical testing of the environmental fate of chemicals and their effect on aquatic species. While such schemes vary in particulars and were designed with different objectives, their design features are much more similar than might be expected. Importantly, most major schemes include the concepts of comparing effect concentrations with exposure concentrations in reaching decisions about the needs for further testing. Essentially all schemes are phased, provide some provision for sequential review and decision, and provide in early stages criteria for such decisions.

Within the evaluative framework, the required or recommended tests vary in species selection and testing details. They also vary in later phases in their recommendations for life-cycle testing. Despite similarity in overall schemes, the differences in test prescription could become burdensome in the international chemical trade.

The recommendation for toxicological testing for the most part focuses on the more established single-species acute, embryo-larvae, and full life-cycle testing for producing or establishing no-effect concentrations. Less familiar tests which involve greater sophistication presently appear less attractive to both chemical industry and the regulator, because many sublethal effects are difficult to evaluate for their environmental significance. While such concentration on established tests seems necessary and appropriate in the catch-up phase of chemicals testing, the research to develop and validate new procedures must be supported. From such research, new tests will emerge to improve our projections from laboratory to field situations, and we could reach decisions with less time and cost. Promising procedures should be explored in later tiers of evaluation schemes where special considerations of chemical structures or exposure will justify their trial and allow validation of experimental procedures. Work for further improvement in the structure of hazard evaluation schemes at present may not be justified. The more pressing needs are to:

(1) establish the meaning of "ecologically significant adverse effect";
(2) establish the approach and the safety factors used to calculate an "acceptable risk";
(3) work for harmonization or cross acceptability of hazard assessment procedures; and
(4) review data accumulating from testing programs to identify those test components of assessment schemes which provide the most cost-effective predictions of risk.

Reference

CAIRNS, J., K. L. DICKSON, AND A. W. MAKI, 1978. Estimating the hazard of chemical substances to aquatic life. ASTM STP 657. Am. Soc. Test. Mater., Philadelphia. 278 pp.

HAZARD ASSESSMENT PHILOSOPHY AND PRINCIPLES

Hazard Assessment Philosophy: A Regulatory Viewpoint

JAMES W. AKERMAN AND DAVID L. COPPAGE

US Environmental Protection Agency
Office of Toxic Substances/Office of Pesticide Programs
401 M Street S.W., Washington, D.C. 20460

Abstract

A number of US Federal laws require aquatic hazard assessment for toxic materials that may enter the aquatic environment. For example, laws administered by the Environmental Protection Agency include the Federal Water Pollution Control Act (FWPCA), the Toxic Substances Control Act (TSCA), and the Federal Insecticide, Fungicide, and Rodenticide Act (FIFRA). The authors' duties are concerned with the implementation of FIFRA and for purposes of this paper, the provisions of aquatic hazard assessment. Relevant portions of FIFRA, regulations for the enforcement of FIFRA, and the registration guidelines are reviewed to indicate what the agency regards as "Criteria for Determinations of Unreasonable Adverse Effects," and what data bases are required to make these determinations. Processes for revision of aquatic hazard assessment are discussed.

Our topic for initiation for discussion concerns regulatory decision criteria for requiring hazard tests, the role of testing and test results in hazard assessment, and when exposure and toxicity findings may indicate unacceptable or unreasonable hazard. Some federal laws administered by the Environmental Protection Agency that require aquatic hazard evaluation of toxic materials are the Federal Water Pollution Control Act (FWPCA), the Toxic Substances Control Act (TSCA), and the Federal Insecticide, Fungicide, and Rodenticide Act (FIFRA). For example, Section 307(a)(2), of FWPCA (P.L. 92-500 of 1972) concerning toxic pollutant effluent standards ("limitations") requires the Environmental Protection Agency (EPA) to collect and consider data relating to "the toxicity of the pollutant, its degradability, the usual or potential presence of the affected organisms in any waters, the importance of the affected organisms, and the nature and extent of the effect of the toxic pollutant on such organisms." Section 307(a)(4), of FWPCA requires the standards or limitations on the effluent provide "an ample margin of safety." Section 2 of TSCA (P.L. 94-469 of October 11, 1976) states "adequate data should be developed with respect to the effect of chemical substances and mixtures on health and the environment. . . ." Section 3 of TSCA states "environment" includes water, air, land, and the interrelationship which exists among and between water, air, and land and all living things. Section 4 of TSCA states "the Administrator shall by rule require that testing be conducted on such substances or mixture to develop data with respect to the health and environmental effects for which there is an insufficiency of data and experience and which are relevant to a determination in the manufacture, distribution in commerce, processing, use, or disposal of such substance or mixture, or that any combination of such activities, does or does not present an unreasonable risk of injury to health or the environment." The FIFRA as amended November 28, 1975 (P.L. 94-140) states in Section 3(c), paragraph 5, that the Administrator shall register a pesticide "when used in accordance with widespread and commonly recognized practice it will not generally cause unreasonable adverse effects on the environment." Section 2(a) of FIFRA states

"The term 'environment' includes water, air, land and all plants and man and other animals living therein, and the interrelationships which exist among these." Section 2(d) of FIFRA states "the term 'animal' means all vertebrate and invertebrate species, including but not limited to man and other mammals, birds, fish and shellfish." Section 3(c), paragraph 2 of FIFRA states "The Administrator shall publish guidelines specifying the kinds of information which will be required to support the registration of a pesticide and shall revise such guidelines from time to time." Since the authors' duties are concerned with FIFRA, we will limit our paper to provisions for aquatic hazard assessment for FIFRA, even though the assessment may apply to other toxic materials.

Discussion

The basic environmental standard for major regulatory determinations under FIFRA, as amended, is "unreasonable adverse effects on the environment." In assessing the aquatic hazards the question arises "What is an unreasonable adverse effect on the organisms in the aquatic environment?" The Regulations for the Enforcement of the FIFRA addresses the above question. General statements are made in Section 162.8 (Data in support of registration and classification) and Section 162.11 (Criteria for determinations of unreasonable adverse effects) which refer to more detailed data requirements to be published in the Registration Guidelines. Section 162.8(b), paragraph 3, states "the applicant shall submit data relative to general and environmental chemistry as specified in the Registration Guidelines." Section 162.8(b), paragraph 4, states "the applicant shall submit data which will be used to assess pesticide hazard to man and to the environment." Laboratory and field studies to make the assessment shall be conducted with either the active ingredient(s) or the specific pesticide formulation or both, as specified in the Registration Guidelines. When data on active ingredient(s) and formulation do not allow a satisfactory basis for decision on product hazard, further studies may also be required to major metabolites, degradation and/or reaction products. The applicant shall submit data obtained through suitable tests which evaluate the hazard to nontarget organisms.

Hazard to Nontarget Organisms

Information of the following types is needed to evaluate nontarget hazards.

(1) Data on acute and subacute toxicity to avian species and acute toxicity to fish and aquatic invertebrates will be required to support the registration of "manufacturing-use products" (such products refer to those which will be reformulated or repackaged).

(2) If the pesticide is intended for outdoor application, data to evaluate hazard to nontarget animals will be required as specified in the Registration Guidelines (outdoor application means any pesticide application or use that occurs outside enclosed man-made structures or consequences of which extend beyond enclosed man-made structures). Conditions under which these data are required depend upon such factors as the pesticide's proposed pattern(s) of use, environmental chemistry characteristics, and nature of the hazard posed to humans and domestic animals. . . . and to other non-target animals. . . . Such data may be obtained from avian reproductive studies, aquatic invertebrate acute toxicity testing, aquatic organism life cycle studies, simulated field testing, and/or field monitoring and observation, as specified in the Registration Guidelines.

(3) If the pesticide is intended for outdoor application and if it is expected to move readily from the application site by means of drift, volatilization, leaching, or lateral movement in soil, then data on toxic effects to susceptible nontarget plants will be required. Section 162.8(d), paragraph 1, states "A registrant or application shall submit any additional data other than that required by these regulations and the Registration Guidelines which the Agency has determined is necessary to support the registration. If any additional information is required on previously registered pesticides, the Agency shall permit sufficient time to obtain such additional information. The Agency shall periodically revise the information needed to support the registration of a pesticide. Such revisions of required information shall be contained in the Registration Guidelines."

Section 162.11 of the Regulations gives a general statement of the "Criteria for Determination of Unreasonable Adverse Effects" as related to the aquatic environment, but also refers many details to the Registration Guidelines. Section 162.11(a), paragraph 3, states "A rebuttable presumption[1] shall arise if a pesticide's ingredient(s), metabolites(s), or degradation product(s) meet or exceed any of the following criteria for risk, as indicated by tests conducted with the animal species and pursuant to the test protocols specified in the Registration Guidelines, or by test results otherwise available. In making this determination, the Agency will take into consideration the type of effect, the statistical significance of the findings and whether the tests were conducted in accordance with the material requirements for valid tests as recognized by experts in the field. . . . ." The major risk criteria stated in Section 162.11 that concern aquatic organisms are: (1) "Results in a maximum calculated concentration following direct application to a 6-inch layer of water more than 1/2 the acute LC50 for aquatic organisms likely to be exposed as measured on test animals specified in the Registration Guidelines"; (2) "Produces any other chronic or delayed toxic effect in test animals at any dosage up to a level, as determined by the Administrator, which is substantially higher than that to which humans can reasonably be anticipated to be exposed, taking into account ample margins of safety"; or (3) "can reasonably be anticipated to result in significant local, regional, or national population reductions in nontarget organisms, or fatality to members of endangered species"; (4) "That, based on toxicological data, epidemiological studies, use history, accident data, monitoring data, or such other evidence as is available to the Administrator, the pesticide poses a substantial question of safety to man or the environment"; or (5) "That the pesticide or its labeling or other material required to be submitted does not comply with the requirements of the Act or when used in accordance with widespread and commonly recognized practice, the pesticide generally causes unreasonable adverse effects on the environment."

Rebuttal of the Administrator's presumption against registration may include proof that when considered with the formulation, packaging, method of use, and widespread and commonly recognized practice of use, the anticipated exposure to an applicator or user and local, regional, or national populations of non-target organisms is not likely to result in any significant acute or subacute adverse effects; or the pesticide will not concentrate, persist, or accrue to levels likely to result in any significant chronic adverse effects; or the risks are outweighed by economic, social, and environmental benefits of use of the pesticide.

It can be seen from the above that an aquatic hazard evaluation program to arrive at decisions regarding the need for further testing depends on the Registration Guidelines. A brief account of the kinds of data required and when they are required by the re-proposed Registration Guidelines follows.

Basic Required Tests

(1) A 96-hour LC50 for one coldwater and one warmwater fish species and the EC50 or LC50 for an aquatic invertebrate are required to support the registration of all manufacturing-use products and all formulated products intended for outdoor application. An exception may be made to this requirement if 30 organisms are exposed to 300,000 times the expected environmental concentration in water or 300 mg/liter (whichever is least) and less than 50% acute mortality occurs. This exception is because we would be assured mortality is unlikely at well above most expected concentrations.

(2) A 96-hour LC50 for shrimp, crab, and fish (marine/estuarine), and either a 48-hour LC50 for embryo-larvae or a 96-hour EC50 shell deposition for molluscs are required to support the registration of a formulated product if the pesticide is intended for direct application to the estuarine or marine environment or if it may be expected to enter this environment in significant concentrations because of its expected use or mobility pattern. Exceptions are as in the preceding paragraph.

Additional Tests

(1) Embryo-larvae or life-cycle tests (or both) of fish and aquatic invertebrates are required to

---

[1] A "presumption" is a notice of intent to deny or cancel registration or to hold a hearing to determine whether the registration should be cancelled or denied upon determination by the Administrator that a pesticide meets or exceeds any of the "risk criteria."

support the registration of a formulated product (the technical grade of each active ingredient in the product shall be tested) if the pesticide product is used in, or is expected to transport to, water from the intended use site, and if the following considerations apply.

Fish embryo-larvae or invertebrate life-cycle tests are indicated:

(a) if the LC50 is less than 1 mg/liter;
(b) if the estimated concentration in water is greater than 0.01 of the LC50;
(c) if actual or estimated concentrations in water resulting from use are less than 0.01 of the LC50 and (i) reproduction is significantly affected as demonstrated by mammalian or avian studies of systems relevant to fish or invertebrates, (ii) physico-chemical properties indicate cumulative effects, or (iii) the pesticide is persistent in water (e.g., half-life in water greater than 4 days);
(d) if the pesticide is intended for broad use such that it is likely to be present in water continuously, regardless of properties.

Fish life-cycle tests are indicated:

(a) if adverse effects are observed in fish embryo-larvae or invertebrate life-cycle test;
(b) if the estimated environmental concentration is one-tenth of the "no-effect" level observed in the fish embryo-larvae or invertebrate life-cycle test;
(c) if significant effects relevant to fish reproduction are observed in mammalian or avian reproduction studies at relatively low concentrations.

(2) Aquatic ecosystem (accumulation) testing is required to support the registration of all formulated products if, after an analysis of the pesticide properties, the individual use patterns, and the results of previous tests, it is determined that use of a pesticide may result in adverse effects on the non-target organisms in aquatic environments, including those of the water column and bottom sediments. Specific requirements will be established on a case-by-case basis. The technical grade of each active ingredient in the product shall be tested.

When a pesticide is used in or is expected to transport to water from the intended use site, major considerations for requiring these toxicity and residue tests include, but are not limited to:

(1) physico-chemical data indicating accumulation (e.g., water solubility less than 0.5 mg/liter and octanol/water partition coefficient greater than 1,000); and
(2) persistence in water (e.g., half-life in water greater than 4 days); or
(3) accumulation in organs and tissues, as indicated by mammalian or avian studies.

A short-term simulated field test (where confined populations are observed), or actual short-term field test (where natural populations are observed), or both, are required if laboratory data indicated adverse short-term acute effects may result from intended use.

A long-term simulated field test (e.g., where reproduction and growth of confined populations are observed) or an actual field test (e.g., where reproduction and growth of natural populations are observed), or both, are required if laboratory data indicate adverse long-term, cumulative, or life-cycle effects may result from intended use. These studies are to be conducted with the formulated product.

Special Test Requirements

In addition to the data required in all succeeding sections of this subpart, data derived from other tests may, under unusual circumstances, be required by the Agency for making judgments regarding safety to aquatic organisms. Such data will be required when special problems are involved, and methods may usually be derived from tests already described or cited in other subparts of the Registration Guidelines. Such data requests may relate to a proposed pattern of use, a toxicological mode of action, or a unique chemical property. The data requested will be specific to the problem. Examples of such test requirements for these unusual circumstances include but are not limited to: certain chemical properties data of Subpart D (§163.61–7); avian acute dermal LD50; eye irritation; acute inhalation LC50; fish or bird cholinesterase tests; metabolism; certain domestic animal safety studies mentioned in §163.86–1; and certain effectiveness data, as required in Subpart G, regarding pest control that may have major impact on the food supply or food chain or a rare or endangered species.

The biological testing decisions in the re-proposed Registration Guidelines depend at several points on physicochemical data such as the expected or estimated concentration in water, life of the pesticide in water, octanol/water partition coefficient, and water solubility. There are Sections in the Registration Guidelines on General Chemistry requirements which are listed in Table 1, and the Environmental Chemistry requirements most of which are listed in Table 2.

Some rationales for the decision criteria are given in a document prepared by the American Institute of Biological Sciences under contract with the Office of Pesticide Programs. This document, entitled "Criteria and Rationale of Decision Making in Aquatic Hazard Evaluation," is found in a Special Technical Publication of the American Society for Testing and Materials (AIBS 1978).

Section 3(d) of the FIFRA states that as part of the registration process the Administrator shall classify the pesticide for general use or restricted use or both. The Regulations under 162.11 provide the classification criteria for general use. Each of the separate criteria must be applied unless the formulation, packaging, or method of use can reasonably be expected to eliminate the route of exposure. The general use classification criteria for aquatic organisms are:

(1) "Application results in a maximum calculated concentration following direct application to a 6-inch layer of water less than 1/10 the acute LC50 for aquatic organisms representative of the organisms likely to be exposed."

(2) The pesticide causes, under conditions of label use, only minor or no discernible adverse effects on the physiology, growth, population levels, or reproductive rates of nontarget organisms, resulting from exposure to the product ingredients, their metabolites, or degradation products.

If the pesticide product use does not meet these criteria, the product is further evaluated according to the Adequacy of Label und Labeling (162.11 (c)(3)). If these criteria are met, the labeling for the affect use will be considered adequate to prevent unreasonable adverse effects on the environment. The labeling criteria include an evaluation of:

(1) the complexity of the required operations or procedures and the need for specialized training or experience;
(2) the widespread and commonly recognized practices of use;
(3) the "safe" margin of error associated with the use;
(4) the need for specialized apparatus normally not available to the general public; and
(5) the effect of failure to follow directions for use in causing delayed or chronic effects.

Congress is currently considering several amendments to FIFRA. One of the significant revisions is the generic standard approach to registering pesticides. This will change our current review procedure from a product-by-product

TABLE 1.—*Summary of physical and chemical properties data requirements in §163.63–8 of the Federal Insecticide, Fungicide, and Rodenticide Act.*

Property	Technical chemical	Manufacturing-use product	Formulated product
Color	X		X
Odor	X		X
Melting point	X		
Solubility	X		
Stability	X		
Octanol/water partition coefficient	X		
Physical state	X	X	X
Density or specific gravity	X	X	X
Boiling point	X	X	X
Vapor pressure	X	X	X
pH	X	X	X
Storage stability		X	X
Flammability		X	X
Oxidizing or reducing action		X	X
Explosiveness		X	X
Miscibility		X	X
Viscosity		X	X
Corrosion characteristics		X	X
Dielectric breakdown voltage			X

TABLE 2.—*Summary of environmental chemistry data requirements by intended use pattern.*[a]

Use patterns	Domestic outdoor	Green-house	Non-crop	Tree fruit/nut crop	Field-vegetable crop	Aquatic food crop	Aquatic non-crop	Forest	Direct discharge	Indirect discharge	Wastewater treatment
Physico-chemical degradation											
Hydrolysis	X	X	X	X	X	X	X	X	X	X	X
Photodegradation		X	X	X	X	X	X	X	X		
Metabolism											
Aerobic soil	X	X	X	X	X			X			
Anaerobic soil					X						
Anaerobic aquatic						X	X	X	X		
Aerobic aquatic						X	X		X		
Effects of microbes on pesticides			X	X	X	X	X	X[b]	X		
Effects of pesticides on microbes			X	X	X	X	X	X	X		
Activated sludge										X	X
Mobility											
Leaching			X	X	X			X[b]			
Volatility		X									
Adsorption	X	X	X	X	X	X	X	X	X		
Water dispersal						X	X		X		
Field dissipation											
Soil	X		X	X	X	X		X			
Water						X	X	X[b]	X		
Ecosystem[b]								X			
Accumulation											
Rotational crop					X	X					
Irrigated crop						X	X				
Fish			X	X	X	X	X	X	X		
Special study							X				

[a] Data requirements cited in § 162.62–10(c)(3), (f), and (g); § 163.62–11(e); § 163.62–12; and § 163.62–13 of the Federal Insecticide, Fungicide, and Rodenticide Act are not included in this table.
[b] Microbe effects on, leaching of, and water transport of chemicals are components of a combined forest ecosystem study.

basis to a broad regulatory decision for an entire group of products that contain the same active ingredient. Other revisions include provisions for conditional registration, waiver of efficacy data requirements, and changes in classification procedure. However, these revisions are not expected to impact on the hazard assessment philosophy as outlined.

Reference

AMERICAN INSTITUTE OF BIOLOGICAL SCIENCE. 1978. Criteria and rationale for decision making in aquatic hazard evaluation. Pages 241–273 *in* J. Cairns, Jr., K. L. Dickson, and A. W. Maki, eds. Estimating the hazard of chemical substances to aquatic life. ASTM STP 657. Am. Soc. Test. Mater., Philadelphia.

An Integrated Approach to Assessing the Potential Impact of Organic Chemicals in the Environment

W. Brock Neely

*Environmental Sciences Research
The Dow Chemical Company
Midland, Michigan 48640*

Abstract

With the diversity of the chemicals that have been made, there is a need for a program to assess the potential environmental hazard resulting from the use of various products. Without such a program, the task of experimentally making an assessment for each material becomes impossible. It will be impossible, for the reason that the manpower and resources available are not equal to the task. Furthermore, as will be shown, the need for performing all conceivable tests is not necessary.

The paper will outline an approach that integrates the basic physical properties of an organic chemical into a cohesive pattern of how the chemical distributes itself in the environment. This profile allows the investigator to visualize the environmental distribution. From such a distribution further environmental tests, if needed, can be designed in a logical and sequential manner.

The proposed decision tree will be examined by applying it to several well-known products including pesticides and industrial chemicals. It will be seen that application of this model in early stage development can aid in the decision making process. By being aware of the potential environmental problems before they occur, greater precautionary measures can be taken in both the manufacture and distribution of the material. Such action will minimize the effect that the planned action might have on the various ecosystems. At the very least, the proposed action can take place with a greater awareness of the potential risk.

The chemical industry has long been concerned with the health and environmental properties of the products that they manufacture and distribute. The effort that is expended in this area has grown exponentially in the past few years due to our growing understanding of the environment. This increased awareness of potential problems is requiring better predictive techniques for making early decisions on what tests are needed (see, for example, Howard et al. 1978). Of necessity, such predictions must be based on laboratory findings, since it is not feasible to use the environment as a testing ground and, in addition, the newly enacted Toxic Substances Control Act (TSCA) requires a company to submit information to the EPA prior to manufacture and distribution.

Section 5 of TSCA, dealing with premanufacture notification, has generated interest in defining the tests that predict the environmental impact of a chemical. One of the concepts that is emerging is based on tier testing (Duthie 1977). The objective of this approach is to enable the studies to proceed in a logical manner and to optimize the amount of information in a cost effective manner. A decision tree will be helpful in this regard, in order to ensure that the critical tests are performed and to prevent needless experimentation. The basic process in any hazard evaluation involving the environmental effects of chemicals is to make predictions of the expected environmental concentration (EEC) and to match this with the experimentally determined no-effect level for appropriate environmental organisms. Once the data demonstrate that the EEC is below the no-effect level, the product should be considered acceptable from an environmental point of view. Estimating environmental exposure is difficult. It may be accomplished for a localized situation where the source inputs and the ecosystem such as a river or lake can be identified. Atmospheric exposures can also be estimated for volatile compounds. However, in most other systems reliance is made on the benchmark approach (Goring 1972). In such an approach the properties of a new chemical are matched with similar chemicals of known environmental distribution, e.g., "DDT-like materials" will behave like DDT.

This paper will present a technique that estimates the distribution of the chemical in the air, water and soil. By comparing this profile with the intended use pattern, decisions can be made on what further action is required. It should be pointed out that this model is designed for assessing environmental as opposed to human health hazard. A different approach will be required for this latter decision.

The discussion will conclude with the presentation of several case studies using existing products.

The Decision Tree

The proposed decision tree (Figure 1) is the extension of several previous studies on compartmental analysis (Blau et al. 1975; Neely and Blau 1977; Neely in press). The output from this tree is a ranking of the environmental distribution to be expected in the three main compartments: air, water, and soil. While the results are given in percents, the numbers are not meant to be absolute but are designed to yield a relative rank of importance. By matching this profile against the use pattern of the chemical it becomes easier to decide on what future tests may be required. The "YES" output from Boxes K, L, and M indicate the need for further studies on degradation.

The lettered boxes in Figure 1 will be described below.

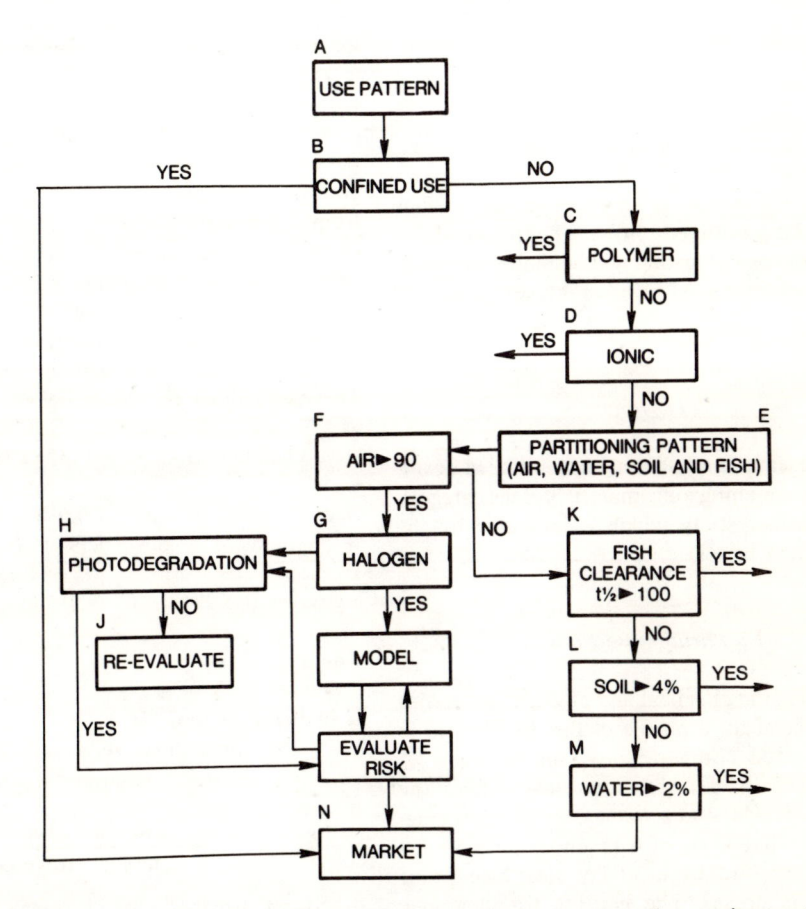

FIGURE 1.—*Schematic representation of a decision tree for designing an appropriate environmental testing program.*

A: Use Pattern

A brief description of the product should be given indicating how the material will be used. For example, if it is a pesticide, it will be intentionally distributed over a wide area; a solvent, on the other hand, will primarily enter the atmosphere. In addition to the use pattern, an estimate of the initial amount and planned rate of environmental entry is required.

B: Confined Use

If the product is used in a manner which prevents entry into the environment, obviously no further testing is required. This is based on the proposition that hazard is a function of both toxicity and exposure. If the environmental concentration is essentially negligible, the hazard will be negligible.

C: Polymer

If the product is an insoluble polymer, e.g., polyethylene, then the environmental considerations will be associated with solid waste. This should be examined; however, the need for further biological testing will be limited to whatever is required in a landfill or incineration operation.

D: Ionic

At present a methodology has not been developed for screening ionic materials. This category needs further study which is indicated by the output from Box D.

E: Partitioning Pattern

By using the chemical and physical properties of the chemical, a profile of the distribution in the indicated compartments can be made. A scenario is used for this purpose where the chemical is added to a water compartment (Figure 2) at a fixed rate of 0.15 g/hour for a 30-day period, followed by a 30-day clearance phase (Neely, in press). The half-life for clearance from the fish biomass is estimated and the percent of the total material found at 30 days in the air, water and soil compartments are calculated.

The estimated half-life ($t_{\frac{1}{2}}$) for clearance from fish is that which would be observed in this ecosystem, which depends on the system parameters (water depth, etc.) and is not to be confused with the clearance rate of a chemical from fish in pure water.

Using a series of common chemicals ranging from toluene to DDT exhibiting a wide range of solubilities and vapor pressures, four regression equations were found to describe the results in a statistically significant manner (Neely 1978a). These equations are shown below:

% of chemical in air = $-0.247\,(1/H)$
 $+ 7.9 \log S + 100.6;$
% of chemical in water = $0.054\,(1/H) + 1.32;$
% of chemical in soil = $0.194\,(1/H)$
 $- 7.65 \log S - 1.93;$
$\log(t_{\frac{1}{2}}) = 0.0027\,(1/H) - 0.282 \log S + 1.08;$

where

H (mm Hg m^3 mole^{-1})
 $= \dfrac{\text{vapor pressure} \times \text{molecular weight}}{\text{solubility (ppm)}};$

S (mM liter^{-1}) $= \dfrac{\text{solubility (ppm)}}{\text{molecular weight}};$

$t_{\frac{1}{2}}$ (hours) = half-life for clearance from fish in this ecosystem.

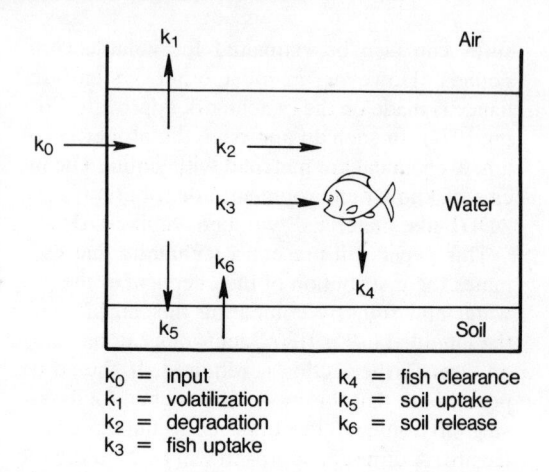

FIGURE 2.—*Compartmental model showing the movement and distribution of a chemical in an aquatic ecosystem.*

k_0 = input
k_1 = volatilization
k_2 = degradation
k_3 = fish uptake
k_4 = fish clearance
k_5 = soil uptake
k_6 = soil release

TABLE 1.—*Properties of a series of chemicals tested in the simulated aquatic ecosystem.*

Chemical	Molecular weight	Vapor pressure (mm Hg)	Water Solubility (ppm)
Toluene	92	30	470
p-Dichlorobenzene	147	1	79
Trichlorobenzene	180	0.5	30
Hexachlorobenzene	285	10^{-5}	0.035
Diphenyl	154	9.7×10^{-3}	7.5
Trichlorobiphenyl	256	1.5×10^{-3}	0.05
Tetrachlorobiphenyl	291	4.9×10^{-4}	0.05
Pentachlorobiphenyl	325	7.7×10^{-5}	0.01
DDT	350	10^{-7}	1.2×10^{-3}
Perchloroethylene	166	14	150

The chemicals along with the relevant data are shown in Table 1. Table 2 shows the results of the computer simulation and the prediction by means of these regression equations.

F: Air > 90%

If the results of the analysis in E indicate that 90% or greater of the chemical is in the air compartment, proceed to G, otherwise proceed to K.

G: Halogen

If the agent contains halogen, an analysis of the mass entering the stratosphere should be conducted. This is based on the theory that chlorine and bromine atoms will cause an increase in the rate at which stratospheric ozone is destroyed (Crutzen et al. 1978; National Academy of Sciences 1976). Models have been proposed to estimate the transfer of chlorine from the troposphere to the stratosphere (Crutzen et al. 1978; Neely 1977b). Any suitable model can be used as long as the same technique is used for all cases. Based on the results of the analysis, decisions can be made about the relative risk. For example, it may be desirable to obtain further data as indicated in Figure 1, with the arrow entering box H from the "Evaluate Risk" box and the information generated recycled back through the model. If the volatile chemical does not contain halogen, proceed to box H.

H: Photodegradation Potential

Those chemicals containing C–H or C=C bonds will be susceptible to hydroxyl radical attack in the troposphere and will be degraded (Darnall et al. 1976). In such a situation, no tropospheric buildup will occur. If the use pattern indicates that no potential problem is perceived in a local environment, go to N.

J: Re-Evaluate

Once a material enters this box on the first pass, a re-evaluation of the properties is necessary. If continued development is warranted, then obviously a search for other degradation pathways must be conducted.

TABLE 2.—*Distribution of the chemicals shown in Table 1 in the various compartments of the simulated ecosystem. Numbers in parentheses were estimated from regression equations.*

Chemical	Water, %	Soil, %	Air, %	$t_{\frac{1}{2}}$ from fish[a], hours
Toluene	0.9 (1.33)	0.4 (~0)	98.6 (~100)	10 (7.6)
p-Dichlorobenzene	1.24 (1.31)	1.28 (0.24)	97.5 (98)	15 (14)
Trichlorophenzene	1.33 (1.34)	2.06 (4.09)	96 (94)	17 (20)
Hexachlorobenzene	3.57 (1.98)	39.4 (31)	56 (68)	162 (164)
Diphenyl	2.27 (1.59)	5.4 (9)	92.2 (89)	27 (29)
Trichlorobiphenyl	1.38 (1.33)	15.2 (26)	83 (71)	96 (134)
Tetrachlorobiphenyl	1.5 (1.34)	17 (27)	81 (71)	104 (139)
Pentachlorobiphenyl	1.5 (1.34)	21 (33)	77 (65)	229 (226)
DDT	1.26 (3.17)	67.5 (46.5)	28 (49)	915 (517)
Perchloroethylene	1 (1.32)	1 (~0)	98 (100)	14 (12)

[a]This is the time for clearance from the fish in the simulated aquatic ecosystem once addition of chemical was terminated.

K: Fish Clearance

If $t_{\frac{1}{2}}$ is greater than 100 hours, a potential problem of bioconcentration is indicated. This is an arbitrary decision and is based on the results of Table 2. Using the benchmark concept (Goring 1972), the chemicals in Table 2 with a $t_{\frac{1}{2}}$ greater than 100 hours are known to have bioconcentration problems; consequently, if the chemical screened has this high a number, it should be examined experimentally for degradability and possibly bioconcentration in aqueous systems. If $t_{\frac{1}{2}}$ is less than 100 hours, proceed to L.

L: Soil

Again, using the benchmark approach, the chemicals in Table 2 suggest that 4% is a reasonable cut-off point. In other words, if the amount of chemical in the soil compartment is greater than 4%, degradation in soil needs to be investigated.

M: Water

In a similar manner, if the amount of chemical in the water compartment is greater than 2%, degradation studies are required.

N: Manufacturing Decision

Based on the use pattern, no long-term environmental problem is anticipated and production of the chemical should proceed. This assumes that the industrial hygiene and animal toxicity studies indicate a reasonable balance in the risk-benefit analysis.

This first cut is designed to give some direction to where further testing is needed. Every case will be slightly different and attempting to formulate a decision tree to steer yourself through the many possibilities would be a wasted exercise. The only firm conclusion is that testing should be continued until enough is known about degradation, distribution, and toxicity of the compound to insure that the expected environmental concentration resulting from the use is below the no effect level. Once this is demonstrated, manufacture and distribution should be allowed.

If in a particular application the concentration reflecting no adverse biological effect is close to the expected environmental level, then more refined measurements on the ecosystem will be required. For example, the actual receiving body of water will need characterization. Some typical properties are shown in Table 3. Simultaneously, an improved estimate of the input function will be needed. Such a function should describe the rate and amount at which the product is anticipated to enter the particular ecosystem.

TABLE 3.—*Typical properties of the aquatic environment needed to predict the concentration of a chemical in that environment.*

Surface area
Depth
pH
Flow/turbulence
% carbon in sediment
Temperature
Salinity
Suspended sediment concentration
Trophic status

Case Studies

Kepone

Kepone is a chemical that has received a great deal of attention (see, for example, Kryielski and Bennett 1978; Dawson et al. 1978). Produced primarily for use as pesticide, it was accidentally discharged into the James River from the man-

TABLE 4.—*Properties of chemicals examined for potential environmental hazard.*

Chemical	Molecular weight	Vapor pressure (mm Hg)	Water Solubility mg/liter
Kepone[a]	491	2.5×10^{-5}	3 at pH 7.0
Mirex[a]	546	6×10^{-6}	0.005
Chlorpyrifos	350	1.9×10^{-5}	2
Monochlorobenzene	112.5	11.8	490

[a]Values obtained from G. Dawson, Battelle Pacific Northwest Laboratories, Richmond, Washington.

ufacturing site at Hopewell, Virginia. The physical properties are listed in Table 4. Following the decision tree in Figure 1, it becomes quickly apparent that Box E is reached. Performing the analysis indicated by Box E, the results in Table 5 are generated. This profile immediately suggests the types of problems that can be associated with the distribution of such a chemical in an aquatic system. These may be listed as follows:

(1) The potential for bioconcentration is evident by the half-life for clearance (greater than 100 hours from the simulated ecosystem).
(2) The great affinity for the soil and water suggests a major problem in these compartments with the continued release of Kepone into an aquatic environment.

This analysis indicates the need for further testing on possible degradative mechanisms. Such tests have been performed and indicated the following: Kepone is persistent in the environment, i.e., it resists photo- and biological degradation and it does in fact bioconcentrate (Dawson et al. 1978). These results do not alter the conclusions from the preliminary analysis.

Furthermore, these conclusions reflect the type of problems that were created by the discharge of the chemical from the manufacturing plant at Hopewell, Virginia (Dawson et al. 1978). Levels ranging up to 10 ppm were found in the James River sediment and high concentrations were found in the Chesapeake Bay (Krygielski and Bennett 1978). Even the ambient air near the plant contained detectable levels of Kepone (Krygielski and Bennett 1978), affirming the predicted release to the atmosphere from the results in Table 2. Dawson et al. (1978) estimated that up to 200,000 pounds of Kepone were released from the Virginia site; furthermore, it is estimated that up to 1/4 of this amount currently resides in the sediments of the river. Thus, it is seen that the actual field observations agree with the profile generated by the equation in Box E of Figure 1 and shown in Table 5.

The examination of the Kepone incident indicates that the proposed decision tree does have the capability of quickly focusing on the key areas for further testing. It also serves as an alert system for what precautions are necessary in both the manufacturing and distribution of the product.

TABLE 5.—*The partitioning pattern generated from box E of the decision tree analysis*[a]

Chemical	% of Chemical in the			$t_{\frac{1}{2}}$ for clearance from fish, hours
	Soil	Air	Water	
Kepone	62	23	14	231
Mirex	37	60	1.4	320
Chlorpyrifos	74	8.5	18	335
Monochlorobenzene	~0	~100	1.34	8

[a]This partitioning is based on physical properties and it does not include any type of degradation mechanism.

Mirex

In 1969, a large-scale, federally coordinated program was implemented to eradicate the imported fire ant in the southeastern United States. The agent chosen for this work was an insecticide known as mirex. While some early warnings over the widespread use of this close relative to Kepone were registered, it was not until mirex was found in fishes from Lake Ontario and in seals from Europe (Kaiser 1978) that the concern over the environmental impact became important. More intensive investigations soon demonstrated that the Lake Ontario ecosystem was badly contaminated. By sampling the bottom sediments of the lake, two distinct sources were apparent; one off the mouth of the Niagara River and the other in the area of Oswego, New York. Since a chemical company on the Niagara River produced mirex, the manufacturing plant was implicated as one of the major sources. The Oswego source was traced back to a plant in Volney, New York.

As the second case study, it is interesting to evaluate mirex through the proposed decision tree in Figure 1. Once again, Box E is reached, using the physical properties of mirex listed in Table 4. The profile of this chlorinated hydrocarbon was determined and is shown in Table 5.

The potential problems associated with mirex become quite evident. The tendency to bioconcentrate in fish is indicated by the long half-life for clearance, while the association with the soil compartment is high. Such a high affinity for sediment suggests that once an aquatic ecosystem becomes contaminated the mirex in the sediment will act as a source for further contamination of the food chain long after the direct source has been terminated. A similar situation has been postulated for the PCB (polychlorinated biphenyl) contamination of Lake Michigan (Neely 1977a). While there are many similarities between mirex and Kepone, there is one important difference (Table 5). In the case of mirex, there is a greater tendency for the chemical to escape into the atmosphere. In many ways mirex more closely resembles DDT. Due to the relatively high volatility rate, both are capable of being circulated around the globe. Fortunately, the production of mirex was much smaller than DDT (50 million pounds of DDT annually at the peak as compared to 50 thousand pounds for mirex) so that detectable levels in species far removed from the source, such as penguins, have not been observed.

However, there is no question that Lake Ontario has become contaminated with mirex. What is important in this discussion is that the simple profile presented in Table 5, combined with further testing showing persistence (Kaiser 1978), has the ability to predict what actually occurred. If such a profile had been generated on a new chemical, the next steps would be to confirm the magnitude of the bioconcentration effect, determine the biodegradation rate in water and soil, and determine the acute and chronic effects on various target organisms. Armed with such information, the producer would be alerted to the dangers of excessive discharges from the manufacturing site. This would allow time to build proper safeguards into the process in order to prevent such an incident from occurring. However, given a proper plant design and trained pesticide operators, there appears to be no environmental reason why such a material cannot be used for the intended purpose of controlling the imported fire ant. In the case of mirex, the human health problems may preclude the safe use of the pesticide (Kaiser 1978).

Chlorpyrifos

The third case study involves chlorpyrifos (O,O-diethyl,3,5,6-trichloro-2-pyridyl phosphorothioate). The key properties are shown in Table 4 and the profile resulting from the application of the equations in Box E of Figure 1 are given in Table 5. Without any further data, the profile suggests similar problems to Kepone. Obviously, before such an insecticide can be widely distributed, degradation studies are needed. Such experiments were performed and indicated a rapid hydrolysis in water (Schaeffer and Dupras 1970), a significant rate of metabolism by fish (Smith et al. 1966), and a rapid destruction by photodegradation in both air and water (Smith 1968). When all of these rate constants were included in the computer simulation (Neely and Blau 1977), a much faster fish clearance time (less than 100 hours) was observed. In addition, the major portion of the added insecticide ended up as hydrolysis products (Neely et al. 1977). Prior experimentation on the fate of the pyridinol entity led to the conclusion that the aquatic plants and microbial population converted this intermediate to CO_2, NH_3 and H_2O (Smith 1966). Such a situation implies that there is no persistence of chlorpyrifos in an aquatic ecosystem. The only precaution that must be observed is that when the pesticide is distributed into water for insect control, the application rate must be adjusted in order that the initial level is below the acute toxicity level for the fish species that might be present. By knowing the physical characteristics of the receiving body of water (see Table 3), the application rate can be adjusted *via* a computer simulation to achieve this safe level (Neely and Blau 1977).

Monochlorobenzene

The fourth example deals with a well known industrial chemical, monochlorobenzene (MCB). Total world dispersive uses for MCB in 1976 was estimated to be 300 million pounds as a solvent and has grown at 5% for the last 30 years.[1] This yields an initial input of 73 million pounds in 1946. Once again Box E in the decision tree of Figure 1 is reached. The environmental profile from this box is shown in Table 5 based on the properties from Table 4. Obviously, this is an example where the potential environmental problems are atmospheric as opposed to aquatic.

---

[1]This estimate was made by the Product Development of the Dow Chemical Company, Midland, MI, and was based on data published in "Synthetic Organic Chemicals, U.S. Production and Sales," US International Trade Commission data for 1950–1975.

The potential for bioconcentration is very low and the amount of material in the soil and water is minimal. From an aquatic long-term environmental point of view, there should be no problems associated with the normal use of MCB in commerce.

The main reason for including this product as a case study is to illustrate the remainder of the decision tree. Studies by Dilling et al. (1976) indicated that monochlorobenzene was rapidly degraded in the troposphere under the influence of radiation from the sun. Using the experimentally derived rate constant of 0.01 hour^{-1} (Dilling et al. 1976),[2] and the release rate into the environment of 73 million pounds in 1946 with a growth constant of 5%, it was possible to estimate the amount of chlorine atom reaching the stratosphere. The model of Neely and Plonka (1978) was used for this estimation and it indicated that 99+% of the chemical was degraded in the lower troposphere. This conclusion indicates that the amount of chlorine entering the stratosphere is negligible and the resultant ozone destruction would be of no consequence. The tropospheric concentration, when photodegradation is incorporated, is less than a ppt (too low to be measured). Providing that the industrial manufacturing and consuming sites meet acceptable water and air quality criteria and good industrial hygiene practices are followed, this is a product that should exhibit no long-term problem.

Conclusion

These four case studies indicate that it is possible to quickly focus in on the key environmental questions that might be associated with a new product. Using the chemical and physical properties, it is possible to visualize where in the environment the chemical will reside. Based on this information, the relevant biological testing can be performed. Incorporating the additional data into the model, a more refined estimate of exposure can be made. Such cycling needs to be performed until the investigator is satisfied that the expected concentration is below the no-effect level. When this is reached, no further testing is required.

---

[2] The reported half-life for MCB was 8.7 hours. This value was corrected to 0.01 hour^{-1} to more closely reflect normal solar radiation.

References

BLAU, G. E., AND W. B. NEELY. 1975. Mathematical model building with an application to determine the distribution of DURSBAN® insecticide added to a simulated ecosystem. Pages 133–163 in A. Macfadyen, ed. Advances in ecological research. Academic Press, New York.

CRUTZEN, P. J., I. A. S. ISAKSEN, AND J. R. MCAFEE. 1978. The impact of the chlorocarbon industry on the ozone layer. J. Geophys. Res. 83:345–363.

DARNALL, K. R., A. C. LLOYD, A. M. WINER, AND J. N. PITTS. 1976. Reactivity scale for atmospheric hydrocarbons based on reaction with hydroxyl radical. Environ. Sci. Technol. 10:692–696.

DAWSON, G. W., J. A. MCNEESE, AND D. C. CHRISTENSEN. 1978. An evaluation of alternatives for the removal/destruction of Kepone residuals in the environment. Proc. 1978 Natl. Conf. Control Hazard. Mater. Spills. April:244–249.

DILLING, W. L., C. J. BREDEWEG, AND N. B. TEFERTILLER. 1976. Simulated atmospheric photodecomposition rates of methylene chloride, trichloroethane, trichloroethylene, and other compounds. Environ. Sci. Technol. 10:351–356.

DUTHIE, J. R. 1977. The importance of sequential assessment in test programs for estimating hazard to aquatic life. Pages 17–35 in F. L. Mayer and J. L. Hamelink, eds. Aquatic toxicology and hazard evaluation. ASTM STP 634. Am. Soc. Test. Mater., Philadelphia.

GORING, C. A. I. 1972. Agricultural chemicals in the environment: a quantitative viewpoint. Pages 793–863 in C. A. I. Gorgin and J. W. Hamaker, eds. Organic chemicals in the soil environment, vol. 2, Marcel Dekker, New York.

HOWARD, P. H., J. SAXENA, AND H. SIKKA. 1978. Determining the fate of chemicals. Environ. Sci. Technol. 12:398–407.

KAISER, K. L. E. 1978. The rise and fall of mirex. Environ. Sci. Technol. 12:520–527.

KRYIELSKI, D. J., AND G. F. BENNETT. 1978. Intergovernmental considerations in the disposal of Kepone—a municipal perspective. Proc. 1978 Natl. Conf. Control Hazard. Mater. Spills. April:274–280.

NATIONAL ACADEMY OF SCIENCES. 1976. Halocarbons: effects on stratospheric ozone. Washington, D.C. 352 pp.

NEELY, W. B. 1977a. A material balance study of polychlorinated biphenyls in Lake Michigan. Sci. Total Environ. 7:117–129.

NEELY, W. B. 1977b. Material balance analysis of trichlorofluoromethane and carbon tetrachloride in the atmosphere. Sci. Total Environ. 8:267–274.

NEELY, W. B. 1978. A method for selecting the most appropriate environmental experiments that need to be performed on a new chemical. Abstr. Am. Chem. Soc. Meet., Div. Environ. Chem., September, 1978.

NEELY, W. B. In press. A preliminary assessment of the environmental exposure to be expected from the addition of a chemical to a simulated aquatic ecosystem. Int. J. Environ. Stud.

NEELY, W. B., AND G. E. BLAU. 1977. The use of laboratory data to predict the distribution of chlorpyrifos in a fish pond. Pages 145–163 *in* M. A. Q. Khan, ed. Pesticides in the aquatic environment. Plenum, New York.

NEELY, W. B., AND J. H. PLONKA. 1978. Estimation of time-averaged hydroxyl radical concentration in the troposphere. Environ. Sci. Technol. 12:317–320.

SHAEFFER, C. H., AND E. F. DUPRAS. 1970. Factors affecting the stability of DURSBAN® in polluted waters. J. Econ. Entomol. 63:701.

SMITH, G. N. 1966. Basic studies on DURSBAN® insecticide. Down Earth 22:3–7.

SMITH, G. N. 1968. Ultraviolet light decomposition studies with DURSBAN® and 3,5,6-trichloro-2-pyridinol. J. Econ. Entomol. 61:793–799.

SMITH, G. N., B. S. WATSON, AND F. S. FISCHER. 1966. The metabolism of ^{14}C-O,O-diethyl O-(3,5,6-trichloro-2-pyridyl) phosophorothioate (DURSBAN®) in fish. J. Econ. Entomol. 59:1464–1475.

An Analysis of Decision Criteria in Environmental Hazard Evaluation Programs

ALAN W. MAKI

Environmental Safety Department
The Procter & Gamble Company
Cincinnati, Ohio 45217

Abstract

Procedures for the formalized evaluation of new and expanded-use chemicals are currently being used by numerous governmental agencies and private groups throughout the world. These programs have resulted in a body of literature typically referred to as hazard evaluation procedures and it is the intent of this paper to review the basic philosophical background and methodology of 13 of these published procedures. Most existing programs demonstrate the application of the tier concept where tests are arranged in definite stages in which the scope and priority of additional testing can be identified. A relatively good agreement exists between the programs with respect to test data requirements in the early tiers; however, a high degree of variation exists in higher level requirements.

Several plans base decisions solely on the fact that a test chemical is proposed for high volume use or wide distribution, regardless of inherent toxicity of the chemical. These are often misleading and tend to specify an inordinate degree of testing that is not scientifically defensible. A comparison of aquatic effects testing data with increasingly refined estimates of the expected environmental concentration of a test substance, LC50/EEC, appears to be the most useful and objective decision criterion used within the schemes to specify need and direction of further testing.

Aquatic toxicology is rapidly emerging as a recognized scientific discipline from a relatively modest beginning approximately 15 years ago. As our understanding of the structure and functioning of the aquatic environment grew, it became obvious that the capacity of our rivers and lakes to assimilate wastes discharged into them was not at all infinite and that serious degradation of water quality was the inevitable result of misuse and mismanagement of this invaluable resource. As techniques of analytical chemistry became increasingly more sophisticated and discerning, it became evident that there remain, in some wastewaters, small concentrations of chemical substances that with inadequate treatment or dilution may significantly impair the survival of resident aquatic life. The persistence and accumulation of undesired toxic substances such as certain pesticides and recalcitrant organics have resulted in the need for preventive measures to ensure protection of aquatic life.

Recognizing the need for additional legislation to control the manufacture and use of thousands of new chemical substances potentially impacting the environment within a year, the US Toxic Substances Control Act (TSCA) was passed in late 1976. This law provides that no person may manufacture a new chemical substance or process a chemical substance for a new use without prenotification of the Environmental Protection Agency. TSCA represents an attempt to establish a mechanism whereby the hazard to human health and the environment from a chemical substance can be assessed before it is introduced into the environment. If the chemical substance presents an unreasonable risk of injury to health or the environment, the Administrator of the EPA may restrict or ban the chemical substance.

The enactment of TSCA has identified the need for consistent, integrated procedures by which the regulator and regulated can ascertain the potential risks to the environment associated with the use of a chemical substance and its potential for release to the environment. To date, these analyses of risk have been somewhat arbitrary with a wide range of laboratory test methods designed to determine biological effect concentrations and estimates of environmental concentrations. These programs have resulted in a body of literature typically referred to as hazard evaluation procedures, and it is the intent of this paper to review the basic philosophical background and methodology of several of these published procedures.

The Hazard Evaluation Process

Basically in any hazard evaluation process, two parallel lines of investigation exist to relate observed biological effects to expected exposure concentration (Figure 1). Theoretically, there exists a concentration of a particular chemical substance which can be determined to have no adverse effects on survival, growth, or reproduction of representative aquatic life. This concentration is typically referred to as the No Observed Effect Concentration (NOEC) and is determined from full-life cycle chronic toxicity testing of fish, macroinvertebrate, or plant species. Similarly, there exists a highest expected environmental concentration (EEC) which will result from the normal anticipated use of the chemical during manufacture, transport, and consumer use. It becomes basic, then, to the hazard evaluation procedure to accurately measure or estimate these two concentrations so that a relative comparison of the two can be made to discern the safety factor or margin of difference between known biological effects and estimated environmental concentration. This figure represents the two concentrations as parallel lines and demonstrates that increasingly more accurate and statistically reliable estimates of these concentrations will result from a sequential series of tests completed in time along the x-axis of the graph. In the early phases of the hazard evaluation process, estimates are made of biological effects and environmental concentration. However, a wide variance exists around these estimates with little confidence; therefore, we see the wide confidence intervals which overlap signifying that additional data are needed to determine if the two concentrations are statistically different. As the hazard evaluation process proceeds, increasingly more accurate estimates of fate and effects can be made to the point where it becomes possible to state with a high degree of confidence that environmental concentrations and biological effect concentrations are indeed different. It then becomes a matter of judgment to determine just how far into the hazard evaluation process the investigator should proceed to further narrow confidence around fate and effect concentrations.

Implied in the time axis of Figure 1 is an integrated evaluation procedure which an investigator can follow to an identifiable end to develop the needed estimates of fate and effects in both freshwater and marine environments. The general trend among investigators has been to categorize the programs into sequential tiers of evaluation, each tier becoming increasingly more complex, time consuming, and expensive with respect to data requirements and testing methodology. The tier concept is extremely useful for the identification of data requirements since it serves to lend a sense of order to the informational requirements and outline the most direct path for accurate estimates of meaningful

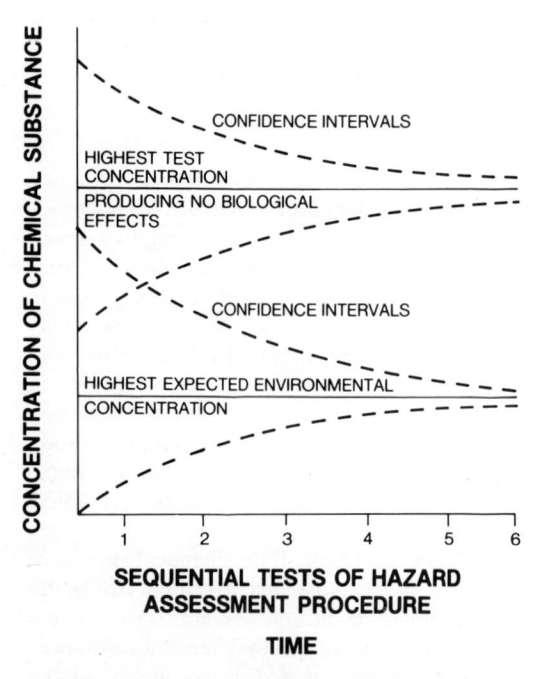

FIGURE 1.—*Diagrammatic representation of a sequential hazard assessment procedure demonstrating increasingly narrow confidence limits for estimates of no-biological-effect concentration and actual expected environmental concentration.*

biological effects and approaches for fate determinations.

To work properly and efficiently, the tier concept in practice must contain an element of specific decision criteria or pass/fail options to provide direction in decisions concerning future use, abandonment, or what additional data may be required in the overall hazard evaluation. Without these decision criteria imposed throughout the hazard evaluation, the entire process degenerates to a simple listing of tests and required data with no real direction or keys identifying when the process has been successfully or satisfactorily completed. It is the intent of this paper to examine the tier concept in use, the identified data requirements, and the decision criteria providing direction to the evaluation process, as exemplified in several published hazard evaluation programs.

It is recognized that numerous other draft programs exist and that this paper is not all inclusive; however, for the sake of clarity and brevity, 13 existing programs were selected as representative of current thinking in this area. This paper is not meant to endorse or detract from any of the hazard programs considered but simply to offer a discussion of each program and examples of how they can be used to reach meaningful decisions regarding the evaluation of new chemical hazards to aquatic life.

Programs for the Evaluation of Hazard

A listing of the hazard evaluation programs considered, respective authors, and purpose of each is presented in Table 1. While it is recognized that many of these programs are designed to include both human and ecological effects testing, the focus of the present consideration will be on testing designed to determine effects on representative aquatic life. Also for the sake of brevity, only a short discussion of the specific decision criteria contained within each testing program is presented here. A full discussion of each program, required testing, and intended application can be found by consulting each of the published references.

AIBS

Authored by a special task group called together by the American Institute of Biological Sciences (AIBS) at the request of EPA, this program is designed to assess basic data submitted for pesticide registration and to specifically indicate the need for and type of additional testing required. Basic information relating to physical/chemical/biological properties of the pesticide and its intended use pattern are used to generate biological response and environmental exposure concentrations. The program presents a numerical ranking scheme based on use pattern to evolve a numerical criterion estimating overall importance of the usage factor to the hazard evaluation procedure.

The program goes further in assigning criteria for 6 basic toxicity and chemical fate tests, any of which denote extra testing considerations (Table 2). The program assumes that at least this base data set considering the six indicated tests

TABLE 1.—*Hazard evaluation schemes.*

Title	Authors	Purpose
AIBS	Ward et al., 1977	Additional testing for pesticide registration
Kodak	Astill et al., 1977	Health & environmental effects
Monsanto	Kimerle et al., 1978	New chemical testing
Unilever	CM Lee, 1978	Aquatic toxicity testing
Stufenplan	German Council Directive, 1977	Health & environmental effects
ASTM	Duthie et al., 1977	Aquatic safety testing
Lloyd	R. Lloyd, 1978	New chemical assessment and water quality criteria
Hueck	H.J. Hueck, 1978	Environmental effects of new chemicals
FIFRA	EPA, 1978	Registering pesticides
Conservation Foundation	A.K. Ahmed et al., 1978	TSCA guidelines
Dow	Dow Chemical Company, 1972	New chemical testing
SDIA	Soap & Detergent Ind. Assoc., 1977	Voluntary notification scheme
TSCA Section 5	EPA/OTS, 1978	Premanufacture notification

TABLE 2.—*AIBS—Criteria and rationale for decision making in aquatic hazard evaluation.*

---

1. Acute toxicity data
    LC50<1.0 mg/liter

2. Estimated environmental concentration
    EEC>0.01 (LC50)

3. Observed mammalian/avian reproductive effects

4. Observed Solubility <0.5 mg/liter
    Octanol/water partitioning >1,000

5. Half-life>4 days

6. Broad, large volume use pattern

---

will be available for all materials and thus allow for the assessment of additional testing requirements (Ward et al. 1978).

Kodak

Eastman Kodak Company has evolved a highly quantified testing program as a means of identifying the magnitude of both human health and ecological exposures and rating them in a numerical context, and by summation, arriving at specified appropriate degree of testing. Four levels of testing information are proposed, each successive tier containing a number of suggested tests which can be chosen to supply necessary information (Table 3). The numerical ratings for each of the criteria are summed and the total rating then determines the additional tiers of testing which must be completed prior to reaching an ultimate decision on the safety of the chemical. All chemicals must be tested through Tier 0 and I and the extent of further testing depends upon the sum of health and ecological

TABLE 3.—*Kodak—tier testing scheme.*

Tier 0	Tier II
Physical/chemical data	90-day feeding studies
Volume use and disposal	Biodegradability
Structure-activity correlates	Algal toxicity
	Bioconcentration
Tier I	Tier III
Basic acute studies	2-year rodent chronic
Irritation and inhalation	Teratology
In vitro mutagenesis	Pharmacokinetics
	Model ecosystem

Quantity discharged	Rating	Biodegradation	Rating
<1 × 10⁵ lbs/year	1	BOD₅/COD >0.5	1
1 × 10⁵ lbs/year			
5 × 10⁶ lbs/year	2	0.25–0.5	2
>5 × 10⁶ lbs/year	3	<0.25	3
Discharge extent		Waste treatment IC50	
One location	1	>5,000 mg/liter	1
Limited	2	250–5,000	2
General	3	<250	3
Octanol/HOH		Fish toxicity	
<100	1	LC50>1,000	1
100–1,000	2	100–1,000	2
>10,000	3	<100	3

Sum of all ratings	Tiers required
< 7	0,I
> 8	0,I,II
>13	0,I,II,III

criteria. While rigid in its specification of additional testing tiers required, the program leaves a considerable degree of latitude within each tier for actual selection of specific tests. The scheme is oriented particularly to the production or processing of new chemicals and its application to the consumer products industry where exposures and volume use considerations are somewhat larger would indicate that testing of all consumer products chemicals would have to proceed through the entire tier evaluation (Astill et al. 1977).

Monsanto

This procedure, presented for discussion purposes only at the 1977 Pellston conference, utilizes a continuous comparison of environmental fate and aquatic organism toxicity data in a sequential four-phase program identifying tests as either Screening, Predictive, Confirmatory, or Monitoring, each implying increasing degrees of sophistication. Following each phase, specific decisions are reached to continue toxicity testing into longer-term programs or additional species, to terminate the project because of unacceptable risk, or to cease testing due to risk being judged acceptable (Table 4). Specific decision criteria are provided within each testing phase and are based on the ratio of acute or chronic toxicity to increasingly refined estimates of environmental concentration of the test material. These specific criteria can be overriden in several instances by less specific criteria such as large-volume use and disposal patterns. Proceeding into the upper phases of testing also introduces a greater degree of needed flexibility with respect to test designs and methodology since techniques of this degree of complexity necessarily must be investigative and specific for the individual test substance (Kimerle et al. 1978).

TABLE 4.—*Monsanto—safety assessment of new chemicals.*

Phase I—Screening
    Chemical/physical characterization
    Biodegradability
    Algal, daphnia and fish toxicity

Phase II—Predictive
    Additional acutes
    Partial chronic
    Full chronic

Phase III—Confirmation
    Model fate studies
    Model ecosystems

Phase IV—Monitoring
    Field studies under actual use

Phase I Screening	Phase II Predictive	Phase III Confirmation	Phase IV Monitoring
LC50/EEC = 1–1000 Large volume, wide use & disposal LC50 <1 mg/liter	LC50/EEC = 1–500 Cumulative toxin High partition coefficient Partial chronic MATC/EEC = 1–50 Large volume use Full chronic MATC/EEC = 1–20	Modeling studies	Field studies

Unilever Research

This program authored by Dr. C. Lee of Unilever's Port Sunlight Laboratory, U.K., is focused primarily on the application and relevance of aquatic toxicity data to a hazard evaluation process. The presentation emphasizes the utility of a well established dose/response relationship to the predictive evaluation of a new chemical's impact. In the context of a hazard evaluation scheme designed to predict the environmental impact of a new chemical, comparisons between estimated environmental concentrations or exposure levels and the LC50 value do provide the basis for specific decision criteria (Table 5). The program demonstrates that acute toxicity values that vary significantly between species, or an apparently slow biodegradation rate, can indicate the need for additional chronic testing. It is also important to note that the prediction of ultimate effects of a chemical on the aquatic environment will rely on many factors, only one of which is inherent toxicity and that each new chemical must therefore be subject to individually designed evaluation procedures under a carefully planned and sequential testing series (Lee 1978).

Stufenplan

The Stufenplan is a German proposal written under Council Directive relating to the testing

TABLE 5.—*Toxicity criteria of Unilever Research (C. M. Lee) and testing tiers of the German Stufenplan.*

Unilever Research—C. M. Lee

Acute toxicity	Decision
EEC <0.01 LC50 for the most sensitive species	
EEC = 0.01–0.1 LC50	No further testing
LC50 data differ by order of magnitude	
Slow biodegradation rate	Chronic tests desirable
EEC = 0.1–1.0 LC50	Additional acute tests; chronic tests
EEC >LC50	Hazard imposed

German Stufenplan

Tonnage individual^{-1} year^{-1}/ total tonnage year^{-1}	Testing tiers	
0/0	Physical/chemical characterization Oral LD50 < 25 mg/kg Cutaneous LD50 < 50 mg/kg Inhalation LD50 < 0.5 mg/liter	Notify authority
10/50	Carcinogenicity/mutagenicity	
100/500	Oral LD50 = 200–2,000 mg/kg Cutaneous LD50 = 400–2,000 mg/kg Inhalation LD50 = 2–20 mg/liter	Notify authority
1,000/5,000	Chronic toxicity Carcinogenicity Embryotoxicity	
2,000/5,000	Fish acute (golden orfe) Environmental behavior Bioaccumulation	

classification, packaging, and labeling of dangerous substances. Of the eleven Articles of the Council Directive, Article 6, Annex A, provides a grading plan which defines the testing obligations of persons placing a new substance on the market in terms of quantity to be used and specific properties of the substance. Quantities of a new chemical requiring governmental notification are expressed as annual tonnage or as the total tonnage per manufacturer placing the substance on the market (Table 5). A flow chart specifying mammalian and aquatic testing requirements indicates the degree of testing required based almost solely on annual tonnage. The new chemical is first subjected to the basic testing series, classified according to results, and tested for specific properties as its use level specifies. As with the Kodak program, the Stufenplan is oriented primarily at production facilities and processes based on annual tonnage. In the consumer products or large chemical industries, the program would specify that the complete testing series be completed due to large volumes used (German Council Directive 1977).

ASTM Subcommittee E-35.21

The American Society for Testing and Materials (ASTM) Committee E-35 on Pesticides, and the Subcommittee E-35.21—Safety to Man and the Environment, have organized a Task Force under the chairmanship of Mr. J. R. Duthie to develop a standard practice for a laboratory testing scheme to evaluate hazard to nontarget aquatic organisms. The current draft of their deliberations is presently in review and as such describes a stepwise laboratory testing scheme to develop data to evaluate the hazard to nontarget aquatic organisms resulting from intended and unintended release of essentially any material into the aquatic environment.

The procedure identifies several key decision points and, depending on resultant data, can consist of as many as three distinct phases: (1) an outline of Priority and Scope, (2) Acute Hazard, and (3) Overall Hazard (Table 6). The testing scheme has provisions for examination of the usage and disposal patterns of a substance and its biological, physical, and toxicological properties in a stepwise manner using sequential evaluation and feedback to select supplementary tests or implement decision alternatives. A number of both objective and subjective decision points are identified where it is appropriate to evaluate all that is known about the material and decide what further information, if any, is needed to adequately assess the environmental hazard of the new or expanded use chemical (Duthie 1977).

Dr. R. Lloyd

Dr. Lloyd's program provides an outline of a progressive, sequential testing program which

TABLE 6.—*Aquatic hazard evaluation scheme: summary of data inputs.*

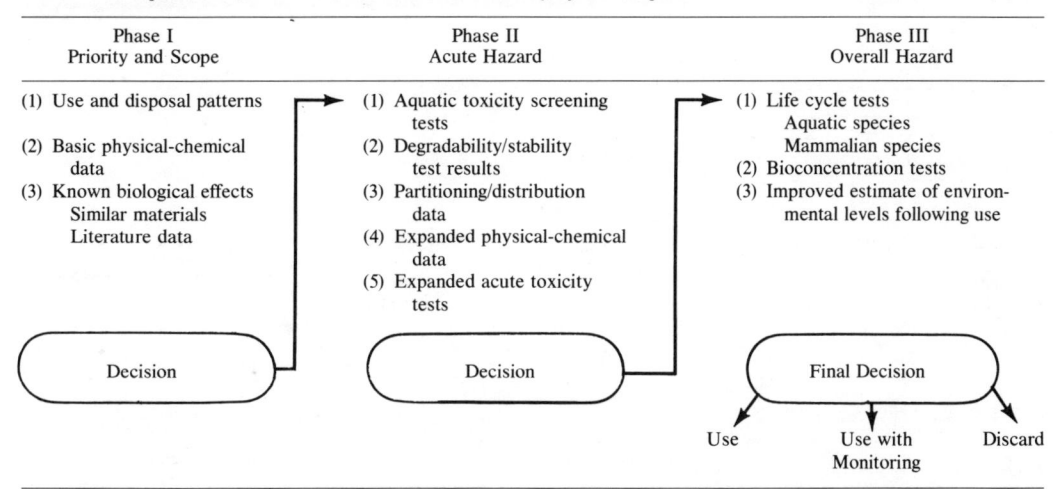

Phase I Priority and Scope	Phase II Acute Hazard	Phase III Overall Hazard
(1) Use and disposal patterns (2) Basic physical-chemical data (3) Known biological effects Similar materials Literature data	(1) Aquatic toxicity screening tests (2) Degradability/stability test results (3) Partitioning/distribution data (4) Expanded physical-chemical data (5) Expanded acute toxicity tests	(1) Life cycle tests Aquatic species Mammalian species (2) Bioconcentration tests (3) Improved estimate of environmental levels following use
Decision	Decision	Final Decision

Final Decision → Use, Use with Monitoring, Discard

can be utilized to evaluate potential risks to aquatic life associated with the release of new or existing chemicals and wastewater effluents. The five-phase program relies heavily upon continual input from environmental fate testing programs to provide information with which to compare the biological effects testing data, thus identifying the need and scope of future testing requirements (Table 7). Provisions are made following each phase to carefully assess the need for further testing. Flexibility in test requirements is specified since a rigid approach would be counter-productive and lead to unnecessary testing and expense. It is stated that the number of instances when testing will be required beyond Phases I and II is likely to be very few. Selected tests within each phase should be chosen and designed to provide for maximum information with ecological relevance and that test objectives and degree of protection of aquatic life have to be defined before testing begins. The program ultimately has direct application to establishment of water quality criteria for the protection of aquatic life (Lloyd 1978).

Dr. H. J. Hueck

The authors present a stepwise testing procedure to provide guidance in the submission of data for establishing water quality criteria and the evaluation of hazard to aquatic life associated with release of new chemicals. A three-phase testing program is outlined which will allow for decisions granting provisional or definite licenses for use (Table 8). The authors recommend that licenses be granted for a limited period of time (5 years) in order to ensure a regular reappraisal of effects following expanded or new uses. It is pointed out by the authors that

TABLE 7.—*Dr. R. Lloyd—sequential testing program.*

Phase I
  Physical/chemical properties
  Estimate of environmental concentration
  Screening tests, mammalian, fish, invertebrate, plant

Phase II
  Improved EEC
  Expanded acute tests
  Stability testing

Phase III
  Partial chronic
  Full chronic
  Organoleptic tests
  Effects of abiotic factors on toxicity

Phase IV
  Bioaccumulation, trophic transport
  Partitioning in soil, air, water, solvents

Phase V
  Field and ecosystem tests

	Phase I	Phase II	Phase III	Phase IV	Phase V
Test data	Acute tests	Expanded acute tests	Partial & full chronic	Bioconcentration	Field tests
Criteria		$MCP = \dfrac{LC50}{AF \times species \times water\ quality}$		$\dfrac{MATC}{EEC} < 10$	
Implication		Phase III or IV testing		Phase V testing	

TABLE 8.—*Dr. H. J. Hueck—stepwise testing procedure for new chemicals.*

Initial Phase
    Physical/chemical data
    Production & use patterns COD, BOD
    Laboratory "die-away"
    30-day bioaccumulation
    Acute tests for 4 trophic levels

Main Phase
    Specific properties
    Mutagenic/carcinogenic effects
    Light & soil interactions
    Metabolite studies
    Model ecosystem food chain test
    Growth, reproduction, and larval tests
    Modeling studies

Confirmatory Phase
    Monitoring in field studies
    Long-term metabolite studies
    Mode-of-action studies
    Long-term mutagenic/teratogenic effects

whereas movement toward standardized testing schemes and methods for analysis of data are desirable, the heterogenity of test methods in many areas of environmental assessment make such rigid standardization improbable.

Test methods used for obtaining aquatic effects data are discussed in some detail and a relative ranking scheme indicating species and test priorities is presented. The scheme emphasizes the utility and importance of mortality, growth, and reproductive effects for the establishment of relevant safety data (Hueck and Hueck-Van der Plas 1975).

FIFRA

Under the Federal Insecticide, Fungicide, and Rodenticide Act (FIFRA), the EPA has published proposed Guidelines for Registering Pesticides which describe specific data requirements and types of tests required for eventual registration of a candidate pesticide. The Guidelines are not really intended as a hazard evaluation procedure but instead serve as a list of test requirements with no real decision criteria as to amount and degree of testing required. They do serve, however, to add further insight into the types of test data considered meaningful when examining effects on aquatic life (Figure 2).

HAZARD EVALUATION TEST SEQUENCE FOR AQUATIC ORGANISMS*

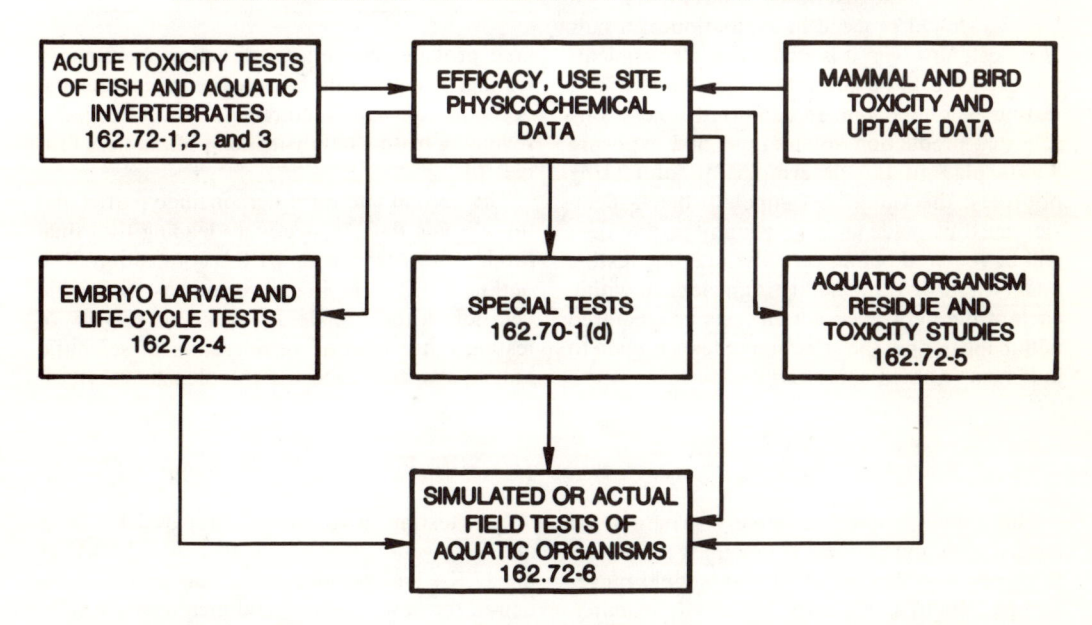

```
ACUTE TOXICITY TESTS          EFFICACY, USE, SITE,          MAMMAL AND BIRD
OF FISH AND AQUATIC    ───▶   PHYSICOCHEMICAL      ◀───    TOXICITY AND
INVERTEBRATES                  DATA                          UPTAKE DATA
162.72-1,2, and 3
                                   │
                                   ▼
EMBRYO LARVAE AND             SPECIAL TESTS                 AQUATIC ORGANISM
LIFE-CYCLE TESTS       ◀──    162.70-1(d)          ───▶    RESIDUE AND
162.72-4                                                    TOXICITY STUDIES
                                                            162.72-5
                                   │
                                   ▼
                              SIMULATED OR ACTUAL
                              FIELD TESTS OF
                              AQUATIC ORGANISMS
                              162.72-6
```

*Numbers in boxes refer to section stating conditions under which testing is required.

FIGURE 2.—*Guidelines for Registering Pesticides—Federal Insecticide, Fungicide, and Rodenticide Act.*

The complex and lengthy document can basically be divided into two distinct testing phases or priorities. Phase 1 contains the specifications for basic chemical/physical characterization along with acute inhalation and irritation studies. The Guidelines offer detailed testing protocols for the conduct of these required tests. Phase II testing direction is specified generally based on results of Phase I tests. The specific numbers and type of additional testing under Phase II are the subject of judgmental decisions and specific application of these Guidelines to broad spectrum or large volume consumer-type products would indicate that all tests listed under Phase I & II should be completed (EPA 1978).

Conservation Foundation–TSCA Guidelines

The Conservation Foundation was contracted by EPA to prepare basic guidelines for testing priorities and requirements under TSCA legislation. The document stresses that only a brief outline of testing requirements is presented with no attempt to provide decision criteria or testing direction. The Guidelines consist of a four-tiered testing system with an increasing degree of complexity (Table 9). No specific criteria are provided to determine just how far into the tier system we should proceed in evaluation of a new chemical; however, it is stated that all chemicals are expected to be subject to Tier 0 and Tier 1 testing as a minimum. In addition to the strong role that production volume, use and exposure levels play in the determination of testing priorities, the Guidelines stipulate that a decision must be made as to how many testing tiers will be required before actually initiating testing under Tier 0 or 1. Flexibility is an element within each of the tiers since not all tests are required within the tier but the selection of tests is open to judgment (Ahmed et al. 1978).

TABLE 9.—*Conservation Foundation—TSCA guidelines.*

Tier 0
- Chemical/physical data
- Analytical technique
- Use & disposal volumes

Tier I
- Acute toxicity screens
- Environmental transformation & degradation
- 10–30 day feeding studies
- 3 in vitro mutagenicity/carcinogenicity tests
- Analytical to 1–10 mg/liter

Tier II
- 90-day subacute tests
- Teratogenic/mutagenic/carcinogenic tests
- Environmental fate
- Bioconcentration tests

Tier III
- Chronic tests
- Multi-generation mutagenicity
- Metabolic studies
- Terrestrial plants & animals

Criteria for number of tiers:
- Production volume
- Use & exposure levels
- Nature of potential hazard
- Physical-chemical properties
- Structure-activity relations

All tested through Tier 0 and I

Dow Chemical

The Dow Chemical Company program for hazard evaluation of new compounds presents a three-phase system initiated with simple range-finding tests to assess acute toxicity to mammalian and aquatic life. Chemical/physical properties and correlations between biological activity and chemical structure to a considerable degree provide the general direction to required testing. The program has provisions for specific decisions on manufacture and use following a review of testing data with each of the tiers (Table 10).

The second and third tiers include partial and full chronic data respectively leading to a final product data review and EPA notification. The outline is very brief in nature and has no specific decision criteria to define scope or priority of testing other than a consideration of scientific judgment within each tier (Moolenaar 1974).

SDIA Voluntary Notification Scheme

This testing program was prepared by Soap and Detergent Industries Association (SIDA) of the U.K., and presents tests and priorities required for new products and ingredients which, in the manufacturer's judgment, may have a significant effect on the environment. It is designed primarily to apply to domestic detergents and

TABLE 10.—*Dow Chemical—methodology for testing new compounds.*

Base data and range finding tests
    Chemical/physical properties
    Oral/cutaneous/inhalation toxicity
    Ames tests, in vitro carcinogenicity
    Acute/partial chronic fish tests
    Bioconcentration estimate
    Biodegradability

First key product review
                      Decision point

Sub-chronic toxicity
30-90-day feeding studies
Skin sensitization

Second key product review
                      Decision point

Advanced studies
Full chronic
Metabolic & pharmacokinetics
Organoleptic, biodegradation

Final key product review
                      Decision point

related products. The testing program is divided into four separate Annexes with a strong emphasis on biodegradability testing and effects on wastewater systems (Table 11). From early information on the use and quantity marketed, it is possible to derive estimates of concentration

TABLE 11.—*SDIA voluntary notification scheme.*

Annex I—Physical/Chemical Properties
    Characterization
    Non-specific analytical method

Annex II—Biodegradability
    Screening
        BOD/CO_2
        Die-away
        Anaerobic
    Model systems
        CAS
        Respirometry

Annex III—Wastewater Treatment Processes
    Digester
    Settling/filtration tests

Annex IV—Sewage Effluent Effects
    Acute fish toxicity
    Phytotoxicity
        WHO mustard seed germination

likely to be found in raw sewage, and a comparison of these data with information concerning sewage works processes allows a prediction of the composition of discharges to the environment which are then compared with biological effects data and toxicity. The program will determine whether all constituents or their degradation products will be present at an environmentally acceptable level following sewage treatment. Where predicted concentrations are low, it is possible to decide on the basis of scientific judgment from limited data that the risk is acceptably small (SDIA 1977).

TCSA Section 5—Premanufacturing Notification

EPA has proposed provisional regulations that govern reporting of new chemical substances under Section 5 of the Toxic Substances Control Act. The testing guidelines specify a base data set of chemical/physical characterizations which must be known for all chemicals and also provide lists of health and ecological effects tests which are similarly required of new chemicals (Table 12).

Flexibility in testing requirements is emphasized in the program to allow manufacturers to rationally allocate scientific resources to assess the introduction into commerce of new chemical substances, taking into account the substances' particular properties, potential toxicological effects, and estimated exposure concentrations. In actuality, the Guidelines leave little room for flexibility in choice of testing requirements with relatively rigid wording specifying completion of all tests listed. Similarly no provisions are available to compare estimated exposure concentrations potentially existing within surface water with laboratory toxicity data to provide a scientific basis for additional test requirements. Based on exposure potential for most materials in the consumer products industry, it appears that all listed test data would be required (EPA Office of Toxic Substances 1978).

Comparative Hazard Evaluations

To further examine the 13 representative hazard evaluation programs and the testing priorities indicated from the specific decision

TABLE 12.—*TSCA Section 5 premanufacturing notifications: physical/chemical properties and environmental fate.*

Base data set
    Spectral data
    Density
    Water solubility
    Octanol/water partition coefficient
    Melting/boiling points
    Dissociation constant for water
    Particle size distribution
    Degradation characteristics
    Flammability
    Vapor pressure
    Adsorption/desorption to particulates
Health effects tests
    Base data set
        Acute toxicity
        Subchronic toxicity
        Teratogenic & reproductive effects
        Mutagenic effects
    Hazard evaluation studies
        Oncogenicity
        Chronic toxicity
        Behavior/neurotoxicity
        Teratogenic/reproductive effects
        Mutagenicity
        Other specific effects
Ecological effects tests
    Microbial—cellulose decomposition, nitrite oxidation, sulfate reduction, respiration
    Algal—freshwater & marine diatom, green & blue-green alga
    Crop plant—seed germination & seedling growth: 3 monocots; 3 dicots
    Fish acute—rainbow trout, bluegill
    Daphnia—acute & chronic: 48-hour & full chronic
    Quail—feeding
    Fish—bioconcentration (as needed)

criteria within each program, comparative hazard evaluations of three hypothetical test chemicals were performed. Realistic testing data were created for the three test materials combining necessary chemical/physical characterizations and fate studies with toxicity data for mammalian and aquatic life (Tables 13–15). Potentials for carcinogenic, teratogenic, and mutagenic effects were specifically eliminated from these considerations in an attempt to simplify the comparative evaluations.

Hypothetical Material 1 was specifically chosen as a relatively non-toxic material with low water solubility and basically little potential for effects on aquatic life. Material 2 was specified as a relatively toxic material, highly water soluble and with an ability to chelate metals in aqueous environments. Material 3 was identified as a highly toxic, intermediately soluble pesticide which degrades relatively rapidly to an even more toxic and persistent intermediate. Relatively complete data files for each of these materials were created and each material was then tested through each of the 13 representative hazard evaluation programs to an identifiable end or until specification for further testing was reached. Data were taken from the specifications for each material as required by the respective programs with the exception of exact chemical formulas for Materials 1, 2, and 3.

The results of this comparative evaluation are summarized in Table 16. Basically as expected, the programs indicate an increasing degree of testing required from the least toxic Material 1 to the highly toxic Material 3 with several schemes specifying a seemingly inordinate degree of testing for even the relatively non-toxic Material 1. Under the AIBS program, a conclusion that some additional acute testing may be required is reached, while Materials 2 and 3 require evaluation through chronic testing, and actual field tests are required for highly toxic Material 3. The rating sums obtained under the Kodak program range from 12 to 14 to 17 for the three materials, indicating that all ecological testing tiers are required for Materials 2 and 3. The summary factor that seems to control the program direction is the wide use and distribution pattern specified for all test materials.

The Monsanto program reaches distinct conclusions for all three materials also. An early decision is reached that Phase 1 screening is all that is required for Material 1, whereas Material 2 must be carried through Phase III confirmatory testing prior to reaching a decision to use the material. An early decision is reached to terminate testing due to unacceptable risks of Material 3 after only acute testing has been completed. Very similar conclusions are reached under the Unilever program indicating only acute tests for Material 1, chronic testing for Material 2, and an indication that a clear hazard was imposed by Material 3.

Although progression through the German Stufenplan is rather rigidly governed by specific decision criteria based solely on use and volume distribution, the conclusion that complete testing would be required for all three test materials is reached based on their relatively large volumes and wide distribution patterns. Under the ASTM program, Material 1 is cleared for use following Phase II testing whereas the additional toxicity imposed by Material 2 implies use only

TABLE 13.—*Hypothetical Test Material 1 used for comparative evaluation of hazard evaluation programs.*

Composition—Pure material, 100% active. No impurities. Exists as a yellow-brown powder. Particle size = 1 μm (range = 0.5–2.0 μm). Specific gravity 1.8. Non-volatile.

Elemental composition—

	SiO_2 = 59%		K = 4%
	Oxides of Fe, Mg = 26%		Ti = 0.5%
	Na = 3%		HOH = 7.5%

Solubility—Extremely low water solubility (<0.1 mg/liter). Exists as a particulate suspension in water. Insoluble in organic solvents.

Analytical method—Extraction, complex formation, and atomic adsorption. Sensitivity from natural waters = 10 mg/liter suspension.

Photodegradation—Half-life under UV light > 10 months. Degrades <5% in 10 mo.

Adsorption to solids—50% of Material 1 found adsorbed to activated sludge solids from a suspension of 100 mg/liter Material 1 and 1,000 mg/liter activated sludge in a completely mixed system.

Aqueous microbial degradation—Slow, ~0.01%/day. Hydrolyzes in basic solutions at pH >8.8 to innocuous materials.

Bioconcentration—BCF <1 for bluegill in flow-through test.

Toxicity—

Acute	96-hour LC50 (mg/liter as suspensions)
Bluegill, *Lepomis macrochirus*	>500 mg/liter
Pink shrimp, *Penaeus duorarum*	~460 mg/liter
Algal toxicity, *Chlorella pyrenoidosa*	96-hour EC50 on growth = 310 mg/liter

Chronic	MATC (mg/liter as suspensions)
Fathead minnow partial chronic	400 > MATC > 320
Daphnia chronic	340 > MATC > 290
Dipteran midge chronic	350 > MATC > 300

Mammalian data—

Species	LD50 (mg/kg)
Rat	>5,500
Rabbit	>7,400

90-day rat feeding study demonstrated histological gut and kidney lesions in individuals fed >200 mg/kg. No changes in blood chemistry at levels <150 mg/kg.

Route of surface water entry—Contained within domestic/industrial sewage effluents.

Estimated sewage effluent concentration—1.8 mg/liter parent molecule as particulate suspension.

---

with field monitoring of fate and effects, and Material 3 is discarded for use based on the acute hazard.

The evaluation program proposed by Dr. R. Lloyd specifies Phase I and II acute testing needed for Material 1 and complete testing through Phase V for Material 2 with the omission of bioconcentration testing. No readily apparent provisions are available to terminate testing in early phases, so Material 3 is carried throughout Phase V with no conclusions regarding ultimate environmental effects. The plan authored by Dr. H. J. Hueck specifies Initial Phase Testing only for Material 1 with Initial and Main Phase Testing required for Material 2. Testing throughout the Confirmatory Phase is indicated for Material 3 with no clear, ultimate decision of environmental hazard.

The FIFRA program for pesticide registration specifies acute and partial chronic testing for Material 1, through simulated field plots for Material 2, and full-scale actual field tests for Material 3. Comparisons of relative hazard to aquatic life are not available for the three materials at the termination of any of the testing programs. The Conservation Foundation program, although possessing a degree of flexibility with respect to test selections, reaches the same conclusions of complete Tier III testing for all three materials based solely on volume use and distribution.

It is possible to infer from the degree of flexibility specified in the Dow Chemical program that testing for Material 1 would be terminated after the completion of the Base Data Set and Initial Range Finding Tests. However, without specific criteria to provide direction, Materials 2 and 3 are tested through Advanced Studies to Final Key Product Review with no clear decision available at that point to judge aquatic hazard. The Soap and Detergent Industries Association program allows for the completion of selected tests only for Material 1 based on the chemical/physical properties but specifies testing through Annex I to IV for Materials 2 and 3.

TABLE 14.—*Hypothetical Test Material 2 used for comparative evaluation of hazard evaluation programs.*

Composition—98% active; exists as a white viscous liquid.
Solubility—Approximately 15 mg/liter in distilled water (parent material). Highly soluble in organic solvents (acetone).
Physical properties—Vapor pressure $<1.3 \times 10^{-5}$ mm; specific gravity = 1.08; boiling point >200°C.
Soil leaching—Intermediate is readily leached from soil by water application. 80% of the degradation intermediate was flushed from a 0.3 m × 10 cm column of brown forest soil after washing with 1 column volume of water.
Degradation—
  Photodegradation: Parent material degrades at 0.5%/day under UV light. Has half-life of 2-3 months. Intermediate non-photolabile.
  Soil degradation: Half-life of parent molecule <5 days. Formation of intermediate observed. After 8 mo 80% of intermediate remains intact. Test done in brown forest loam.
  Aqueous degradation: Intermediate stable in water also. 70% of intermediate remains following 10 mo incubation in natural surface water static system.
Analytical method—Solvent extract, complex formation, and TLC analysis. Sensitivity = 50 ppb for parent molecule and stable intermediate from surface water samples.
Toxicity—

	96-hour LC50 (mg/liter)	
Acute	Parent	Intermediate
Bluegill, *Lepomis macrochirus*	0.91 (0.80–1.02)	0.24 (0.18–0.30)
Pink shrimp, *Penaeus duorarum*	0.79 (0.68–0.90)	0.13 (0.07–0.19)
Algal toxicity, *Scenedesmus quadricauda*	96-hour EC50 on growth = 0.9 mg/liter	

Chronic	MATC (µg/liter)
Fathead minnow full chronic	80 > MATC > 65
Daphnia chronic	50 > MATC > 40
Brook trout partial chronic	80 > MATC > 55

Addition of river bottom sediments to bluegill static acute test reduced toxicity of parent material and intermediate by a factor of 100 by settling the adsorbed compound to the bottom of the chamber.
Partition coefficient—200 for parent molecule; 180 for stable intermediate.
Bioconcentration factor—In bluegill continuous-flow test = 220 (stable intermediate). Uptake rate, K_1(hour^{-1}) = 5.8. Clearance rate, K_2 (hour^{-1}) = 0.025. BCF, K_1/K_2 = 220.
Mammalian and avian data—

	LD50 (mg/kg)	
Species	Parent	Intermediate
Rat	340	91
Bobwhite quail	400	120–130
Mourning dove	280	60–100

Chronic data: Dogs fed 4 mg/kg of the stable intermediate daily for 90 days showed loss of weight, abnormal behavior. Necropsy indicated increased urea content in blood serum, decrease in vitamin C levels in organs. Histological lesions in liver gut epithelium and kidneys.
Route of surface water entry—Contained in land runoff from agricultural applications. Recommended application = 5 lbs/acre. Sprayed on soil in early spring and late summer.
Usage—Pest control agent for major US crops, inland and coastal areas.

Testing guidelines under the Toxic Substances Control Act Section 5 indicate that all Base Data and Health and Ecological Effects testing is required for all three materials based primarily on the volume use and distribution of each material.

Evaluation of Decision Criteria

The comparative evaluation of the 13 individual hazard evaluation programs demonstrates that the conclusions reached with three hypothetical case study materials can vary considerably with respect to amount of testing indicated and final evaluations reached after indicated testing. Although several of the plans indicate a desire to retain a high degree of flexibility in test choices, relatively rigid specifications of high volume use or potential for exposure of large-area environments key the decision to require advanced level or higher tier testing regardless of the inherent lack of toxicity of the test substance.

An examination of the specific criteria contained in the early phases of these programs also demonstrates a wide degree of latitude with re-

TABLE 15.—*Hypothetical Test Material 3 used for comparative evaluation of hazard evaluation programs.*

Composition—97% active, 3% impurities. Exists as a crystalline solid, pale blue in color.
Elemental composition—Carbon, hydrogen, oxygen.
Solubility—Highly water-soluble (8,000 mg/liter).
Analytical method—Solvent extract, complexation, and colorimetric analysis. Sensitivity from natural waters ~0.5 mg/liter.
Photodegradation—Half-life of 25 days under UV light.
Adsorption to solids—Highly adsorbed to activated sludge suspended solids. Freundlich equation $A = KD^n$, where $K = 4.5$, $n = 0.84$.
Aqueous biodegradation—Half-life in natural river water < 4 days. No efforts made to characterize bacterial population of water sample. Degrades to intermediates with toxic effects only at concentrations greater than 100 mg/liter. No persistent intermediates measured.
Metals interaction—Experiments demonstrated the material had a limited capacity to form complexes in aqueous solutions of heavy metals. The formation constants (K) were measured for the following:

Metal	Metal/Material 2 formation constant (K)
Mg^{++}	$10^{2.39}$
Ca^{++}	$10^{1.98}$
Fe^{++}	$10^{5.37}$
Zn^{++}	$10^{5.49}$
Cu^{++}	$10^{7.03}$

Fish bioconcentration—Flow-through test with bluegill: residue in fish <10× aquatic concentration. Octanol/water partition coefficient = 85.
Toxicity—

Acute	96-hour LC50 (mg/liter)
Bluegill, *Lepomis macrochirus*	4.50 (4.28–4.72)
Pink shrimp, *Penaeus duorarum*	3.76 (3.62–3.91)
Algal toxicity, *Scenedesmus quadricauda*	96-hour EC50 on growth = 2.50

Chronic	MATC (mg/liter)
Fathead minnow partial chronic	2.0 > MATC > 1.5
Daphnia chronic (2-generation)	1.75 > MATC > 1.0

Mammalian data—

Species	LD50 mg/kg
Rat	300
Guinea pig	274

Chronic effects: 180-day rabbit feeding study indicated an inhibited excretory hepatic function, change in blood sugar levels, and pathological changes in kidneys and liver at 2.0 mg/kg.
Route of surface water entry—Contained within domestic/industrial sewage effluents.
Estimated maximum sewage effluent concentration—2.2 mg/liter parent molecule.

spect to implications from acute toxicity testing (Table 17). To reach a decision of no further testing required, either a high specific LC50 is observed or the LC50 is specified not to exceed a fractional percentage of the estimated environmental concentration (EEC). These criteria have been taken out of context from one or more of the hazard evaluation programs examined and are compared here to demonstrate the wide variation accompanying the interpretation of acute toxicity data.

Alternatively, to reach the decision that further testing is required, many more factors can be employed. Among them are comparisons with EEC, a high volume use, variable toxicity data among species, a demonstrated cumulative toxin, or reproductive effects from mammalian testing programs (Table 17). One or several of these are listed in the more objective hazard evaluation plans and a positive reading can direct the evaluation to upper level or chronic testing. It is readily obvious that both decisions of further, and no further, testing required are subject to a wide degree of disagreement among the existing programs and that until a larger comparative data base is available for many more real-world test chemicals, specific resolution of the most ecologically relevant or appropriate decision criteria is not likely.

Although a high degree of variation does exist among the testing and points and conclusions of the existing plans, there is still a considerable

TABLE 16.—*Conclusions reached on Hypothetical Materials I-III by various hazard evaluation schemes.*

Plan	Hypothetical Material I	Hypothetical Material II	Hypothetical Material III
AIBS	Possible additional acute testing	Partial & full fish chronics	Full chronics, BCF, field tests needed
Kodak	Rating sum = 12, Tiers 0, I, II	Rating sum = 14, all tiers needed	Rating sum = 17, all tiers required
Monsanto	Phase I screening only	Testing through Phase III confirmation	Terminate testing after acute tests
Unilever	Acute screening tests only	Chronic testing required	Clear hazard imposed
Stufenplan	Complete testing based on use & volume	Complete testing based on use & volume	Complete testing based on use & volume
ASTM	Use following Phase II testing	Use material with monitoring	Discard use based on acute hazard
Lloyd	Phase I and II acute testing	Through Phase V field tests: omit BCF	Complete Phase I–V
Hueck	Initial phase only	Initial & main phase	Test through confirmatory phase
FIFRA	Acute testing & partial chronics	Simulated field testing	Actual field testing
Conservation Foundation	Through tier III based on volume & use	Through Tier III based on volume & use	Through Tier III based on volume & use
Dow	Base data & range finding tests	Advanced studies to final key product	Final key product review
SDIA	Annex I–IV selected tests only	Annex I–IV	Annex I–IV
TSCA § 5	Base data, health & ecological effects	Base data, health & ecological effects	Base data, health & ecological effects

TABLE 17.—*Implications from acute testing. EEC = Estimated Environmental Concentration.*

No further testing required
    LC50 >1,000 mg/liter
    EEC >0.002(LC50)
    EEC <0.01(LC50)
    LC50 ≥100(EEC)
    LC50 >5,000 mg/liter

Further testing required
    LC50/EEC = 1–1,000
    Use/volume/disposal patterns cause unusual concern
    LC50 <1 mg/liter
    LC50 highly variable among species
    Material is a cumulative toxin
    LC50 ≤10(EEC)
    EEC >0.01(LC50)
    Key effects in avian or mammalian efficacy tests
    LC50 <5 mg/liter

degree of basic agreement as to the acquisition of test data among the programs. Each of the programs begins with some requirement for basic chemical/physical characterization data, estimates of use, and volume distribution. Also in the early tiers, measures of acute toxicity to aquatic life are generally required before any relevant decisions can be made. A high degree of variation then follows among the schemes with respect to exactly what data are to be required, in what specific order, and what conclusions can be drawn. Each of the schemes recognizes that fully objective yes and no decisions regarding the need for further testing are difficult to make and as a consequence, the degree of objectivity versus subjectivity in the decision making proc-

```
PLAN
 (1) AIBS                    OBJECTIVE, CLEAR DECISION CRITERIA
 (2) KODAK                              ▲
 (3) MONSANTO                           |
 (4) UNILEVER                           |
 (5) STUFENPLAN                         |
 (6) ASTM                               |
 (7) LLOYD                              |
 (8) HUECK                              |
 (9) FIFRA                              |
(10) CONSERVATION FOUNDATION            |
(11) DOW                                |
(12) SDIA                               ▼
(13) TSCA SECT. 5            SUBJECTIVE, NO CLEAR DECISION CRITERIA
```

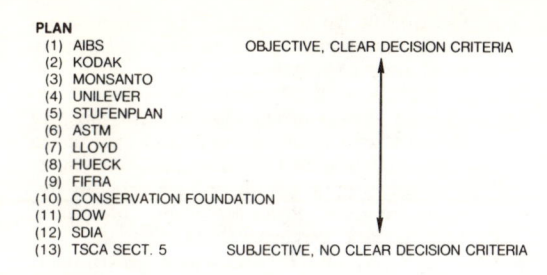

FIGURE 3.—*Decision criteria ranking of hazard evaluation programs.*

TABLE 18.—*Important parameters of hazard evaluation programs for new and expanded-use chemicals.*

Programs of aquatic hazard assessment

Design individuality
Consider usage and disposal patterns
Evaluate chemical and physical properties
Select toxicological tests
Relate effects to concentrations
Provide sequential assessment
Make decisions as early as possible

ess ranges widely between the various plans examined. A general comparison of this degree of objectivity is presented in Figure 3 and ranges from the strictly objective AIBS program for pesticide evaluation to the relatively subjective FIFRA Guidelines. For all basic purposes, no real difference in degree of objectivity can be seen between the last five programs compared on the scale.

Conclusions from the use and application of these hazard evaluation programs to three hypothetical test materials lead to the identification of several desirable factors from a hazard evaluation program (Table 18). To ensure meaningful application of a hazard evaluation program to test substances with a wide variety of physical, chemical and biological properties, the program should not be overly structured or rigid, but instead provide an opportunity for the investigator to design individuality into the testing sequence. Specific use and disposal patterns must be considered early in the program along with an evaluation of chemical and physical properties. These data will yield early identification of the potential environmental distribution and fate of the material which in turn will allow the investigator to identify appropriate test species and conditions of exposure for toxicological testing. Subsequent comparison of the environmental chemistry and fate data with the biological effects data should then define the magnitude of testing required in a sequential assessment procedure and allow the investigator to make decisions as early as possible.

Conclusions

(1) The development of objective and meaningful hazard evaluation programs for test substances potentially reaching the aquatic environment has come far with a relatively high degree of agreement in testing methods and procedures within the relatively short history of aquatic toxicology.

(2) Most existing programs demonstrate the application of the tier concept where tests are arranged in definite stages so that decisions can be made at the end of each stage as to scope and priority of future required testing.

(3) Decision criteria in hazard evaluation programs can be identified as belonging to two basic types: (1) those specifying the direction and/or need for future tests; and (2) those leading to conclusions of degree of risk following a hazard evaluation.

(4) Although a relatively good agreement exists among the 13 programs examined with respect to test data requirements in the initial or early tiers, a high degree of variation exists in second level and higher testing requirements.

(5) The results of these variations in testing requirements were evident in the variable conclusions reached during the evaluation of three hypothetical test chemicals by all 13 programs.

(6) A comparison of aquatic effects testing with increasingly accurate estimates of the expected environmental concentration of a test substance, LC50/EEC, appear to be the most common and objective decision criteria used within the schemes to specify need and direction of further testing.

(7) A greater degree of feedback is needed between biological effects testing and testing designed to discern chemical fate of a material in order to ascertain that effects testing is being done with chemical forms of the test material that potentially could exist in surface waters.

(8) Decisions based solely on the fact that a test chemical is proposed for high volume use or

wide distribution, regardless of inherent toxicity of the chemical, are often misleading and tend to specify an inordinate degree of testing that is not scientifically defensible.

(9) There is a wide degree of disagreement among existing hazard evaluation programs with respect to decision criteria specifying additional test data. Until a larger comparative data base is available for many more real-world test chemicals, specific resolution of the most ecologically relevant or appropriate decision criteria is unlikely.

References

AHMED, A. K. ET AL. 1978. Conservation Foundation Panel, "Approaches for the development of testing guidelines under the Toxic Substances Control Act." Draft 2nd Review 2/21/78.

ASTILL, B. D. ET AL. 1977. "A tier testing scheme." Eastman Kodak Company.

DUTHIE, J. R. 1977. The importance of sequential assessment in test programs for estimating hazard to aquatic life. Pages 17—35 in Aquatic toxicology and hazard evaluation. ASTM STP 634. Am. Soc. Test. Mater., Philadelphia.

EPA (ENVIRONMENTAL PROTECTION AGENCY). 1978. Proposed Guidelines for Registration of Pesticides in the United States. Fed. Regist., vol. 43, #132, July 10, 1978, Part II: 29696–29741.

EPA (ENVIRONMENTAL PROTECTION AGENCY) OFFICE OF TOXIC SUBSTANCES. 1978. Preliminary draft guidance for premanufacture notification, Toxic Substances Control Act Section 5. July 14, 1978.

GERMAN COUNCIL DIRECTIVE. 1977. German proposal for the 6th amendment of the Council Directive of 27 June 1967, Articles 1–22.

HUECK, H. J. AND E. H. HUECK-VAN DER PLAS. 1975. A system of tests for the assessment of potential effects of chemicals and data in the aquatic environment. In Principles and methods of determining ecological criteria on hydrobiocenoses. Proc. Environ. Coloquium, Luxembourg. Pergamon Press.

KIMERLE, R. A., W. E. GLEDHILL, AND G. J. LEVINSKAS. 1978. Environmental safety assessment of new materials. Pages 136–146 in J. Cairns, K. Dickson, and A. Maki, eds. Estimating the hazard of chemical substances to aquatic life. ASTM STP 657. Am. Soc. Test. Mater., Philadelphia.

LEE, C. M. 1978. Aquatic toxicity testing and its relevance to assessment of the environmental acceptability of chemicals. 2nd draft, AIS Working Group: Safety of the Aquatic Environment. April, 1978.

LLOYD, R. 1978. An outline scheme for evaluating the hazard to aquatic life from chemical pollution and obtaining water quality criteria. Sixth FAO/SIDA workshop on aquatic pollution in relation to protection of living resources. June, 1978.

MOOLENAAR, R. 1974. Environment input of chemicals. Seminar on early warning systems for toxic substances, January 1974. EPA-560/1-73-003: 167–174.

SIDA (SOAP AND DETERGENT INDUSTRIES ASSOCIATION). 1977. Voluntary notification scheme. Standing Technical Committee on Synthetic Detergents. November 1977.

WARD, C. H. ET AL. 1978. Criteria and rationale for decision making in aquatic hazard evaluation. AIBS Aquatic Hazard of Pesticides Task Group, for USEPA/OPP Criteria and Evaluation Division (draft, April 1978).

Aquatic Test Organisms and Methods Useful for Assessment of Chronic Toxicity of Chemicals

EUGENE F. KENAGA

*Health and Environmental Research
Dow Chemical USA
Midland, Michigan 48640*

Abstract

A review of literature data on the toxicity of chemicals to fish, a daphnid, and a mammal was made to obtain a basis for efficiently determining chronic no-effect levels for many chemicals as may be required by law. The principal measurements used in statistically determining chronic no-effect levels for chemicals are the acute LC50, the maximum acceptable toxicant concentration (MATC), and the application factor (AF). Equations for these measurements were developed to help in predictive calculation of AF's, MATC's and LC50's for combinations of related and unrelated organisms (fish and *Daphnia*) and chemicals. None of these methods are entirely reliable but are useful for early assessment of chronic toxicity to determine whether or not extensive testing appears necessary. The degree of predictability increases with the use of shortened or short-life-cycle chronic toxicity data.

The enactment of the Toxic Substances Control Act (TSCA), P.L. 94469, on October 11, 1976 provided the US Environmental Protection Agency with the authority to require essential and critically needed pre-market testing for indications of human health and environmental effects of chemical substances. Successful implementation of TSCA may depend on the judicious use of pre-market toxicity screening and evaluation processes.

Selection of a minimal number of good indicator species, representative of the many organisms on Earth, is a basic practical need for use in acute toxicity screening and evaluation of chemicals. A correlation study of eight animal organisms and 75 pesticides was conducted by Kenaga (in press). From this study, rat LD50, fish LC50, and *Daphnia* LC50 values appeared to be the most useful animal test indicators. Equations were developed for the best organism correlations to be used for prediction of acute toxicity for related organisms or with related chemicals.

Another important need for evaluation of toxicity is a method of prediction of chronic effects of chemicals based on as little data as are needed for accuracy and usefulness.

The relationship of chronic and acute toxicity for rats was examined by Kenaga (in press) based on 69 pesticides for which acute or oral LD50 and chronic dietary no-effect (NE) levels were available. There was no good correlation for all compounds. Absence of correlation was probably due to the lack of a common definition of no-effect level for chronic studies, i.e., no-effect level could be based on enzymatic, tissue pathology, reproductive effects, etc., but not mortality, in contrast to the acute oral LD50 value. The rat chronic dietary no-effect values were also compared with bobwhite dietary LC50 values and found to be more closely correlated to them than to the rat oral LD50's, thus emphasizing the importance of using comparable test methods for prediction.

Unfortunately, for correlation purposes, there are little comparable acute-chronic toxicity data available on the same chemical or species for birds, other vertebrates, and invertebrates. The best source of such comparative data involves some recently available information concerning the toxicity of several chemicals to some fish species and invertebrates, mainly *Daphnia magna*. The use of fish toxicity data has been valuable in trying to determine hazards to aquatic organisms. One such hazard determination value is known as the "maximum acceptable toxicant concentration" (MATC). The MATC is essentially a chronic no-effect level that has been determined experimentally on at least one life cycle of an organism and may be used for estimation of MATC's for other related species. Another hazard determination value is known as an "application factor" (AF). The AF is a ratio of the MATC to the acute LC50 value which has been determined for a given species. The AF is used to obtain an estimated MATC value for other species based on a known acute toxicity value (Mount and Stephan 1967).

Since 1967 a considerable amount of fish and daphnid acute and chronic toxicity data has been generated. Mount (1977) has discussed a more recent evaluation of MATC's and AF's and concludes that "The present need for establishing biologically acceptable concentrations of as many as 1500 new products each year, makes crystal clear that our past pace of data generation will have to be increased two to three orders of magnitude. Either more resources must be obtained or else a faster means to produce data must be found. Probably no method will always be correct, and we may have to be content with being right 'most of the time.' Perhaps never before have we faced a challenge so important to our national welfare as the one produced by the information needs of the Toxic Substances Control Act. Since the consequences of being unnecessarily restrictive are different, but perhaps as severe as being too liberal, our best effort will be none too good."

This paper contains a review of the present usefulness of MATC and AF values and their relationship to fish and daphnids.

This paper also contains a summary of data on compounds for which flowing-water LC50 values and prolonged exposure tests are available on the same chemical and organism. Application factors are thus calculated for the data. Statistical evaluation is applied to determine the limits of usefulness of correlations of chronic toxicity data between different species, chemicals, and test methods.

Test Organisms

Certain organisms have been used sufficiently in acute and chronic toxicity tests to support a common data base for the different chemicals discussed in this paper. Many other organisms may do equally well given time for development of the test methods. The test organisms used here are as follows: Salmonidae—rainbow trout (*Salmo gairdneri*), brook trout (*Savelinus fontinalis*), and lake trout (*Salvelinus namaycush*); Centrachidae—bluegill (*Lepomis macrochirus*); Percidae—yellow perch (*Perca flavenscens*); Cyprinidae—fathead minnow (*Pimephales promelas*); Ictaluridae—channel catfish (*Ictalurus punctatus*); Cyprinodontidae—flagfish (*Jordanella floridae*) and sheepshead minnow (*Cypridodon variegatus*); and Daphnidae—water flea (*Daphnia magna*).

Test Methods

Acute Toxicity

A flow-through water exposure technique was used with juvenile fish to determine a 96-hour LC50 or a lethal threshold limit (Adelman et al. 1976; APHA et al. 1975; Hansen and Parrish 1977; Macek et al. 1976a) except for a few cases where a static water test was used as indicated in Table 1. The general procedures were like those described by APHA et al. (1975).

Daphnia acute toxicity tests were 48-hour LC50 determinations conducted in static water (Biesinger and Christensen 1972; Cardwell et al. 1977; Macek et al. 1976a, 1976b, 1976c; Mayer et al. 1977; Pearson et al. 1979).

Chronic Toxicity

Chronic toxicity tests methods for fish involved flow-through systems as conducted by various researchers (Adelman et al. 1976; Allison and Hermanutz 1977; Andrew et al. 1978; Benoit 1975; Biesinger and Christensen 1972; Brungs 1969; Cardwell et al. 1977; Carlson 1971; Eaton 1970, 1974; Hansen and Parrish 1977; Hermanutz 1977; Hermanutz et al. 1973; Jarvinen et al. 1976; Macek et al. 1976a, 1976b, 1976c; Macek and Sleight 1977; Maki 1977; Mayer et al. 1975 and 1977; McKim 1977; McKim and Benoit 1971; Merna and Eisele 1973; Mount and Stephan 1967, 1969; Nebeker et al. 1974; Parrish et al. 1977, 1978; Pickering 1974; Pickering and Gast 1972; Pickering and Thatcher 1970; Woodward 1976). Test methods used for *Daphnia* were those used by the following researchers: Biesinger and Christensen (1972); Cardwell et al. (1977); Macek et al. (1976a, 1976b, 1976c); Maki (1977).

Use of Indicators of Chronic Toxicity for Fish

Acute LC50 Values

As explained by Mount (1977) the LC50 multiplied by an estimated application factor of 0.01 is sometimes used to calculate a safe exposure level for fish in the absence of chronic toxicity testing. As seen by inspection of modern data, this AF as experimentally derived can vary from more than 0.1 to about 0.001 and thus further data are needed for accurate estimation of safe exposure levels of MATC's.

MATC Values

In this paper, chronic toxicity includes MATC data from partial (P) and complete (C) life-cycle tests with fish (or *Daphnia*) involving all developmental stages. This use of data is based on the compilation of McKim (1977) who compared 56 fish life-cycle toxicity tests involving 34 chemicals and four species of fish. The embryo-larval and early juvenile life stages (partial life cycle) were found to be within about 2-fold differences of being the most sensitive to each chemical and can be used to estimate the MATC on the same chemical without complete life-cycle tests previously felt necessary for determining MATC's.

Macek and Sleight (1977) also found that much of the available data indicate that the critical life stage exposure data "provide estimates of chemically safe concentrations (MATC's) remarkably similar to those empirically derived from definitive chronic toxicity studies." This was based on comparison of critical life stages and chronic studies for 22 chemicals.

Availability of suitable chronic toxicity data on fish have been discussed by Andrew et al. (1978). The comparability of literature data leaves much to be desired because of variability in static versus flowing-water test methods, species constancy, use of partial life-cycle tests versus full life-cycle tests, etc. Nevertheless with ten comparable toxicants, they found that a statistically significant mathematical relationship exists between LC50 and MATC values. However, there was no consistent agreement of AF values and MATC values from one toxicant to another (Macek et al. 1976a).

Andrew et al. (1978) developed an empirically derived equation from 31 sets of observed LC50 and MATC values for ten toxicants and five species of fish which fit within a factor of six (less than one order of magnitude). Recommendations were made for greater test method uniformity, a modified application factor, and use of test parameters more sensitive than acute mortality. The authors stated that "Embryo-larval tests and possibly cough tests, should produce useful information about the effects of water quality on MATC values at a reasonable cost since they are usually more sensitive than acute mortality tests."

The ultimate need in hazard evaluation is a no-effect level for each chemical on the life cycle of every organism. Practically, this need is presently met by use of selected indicator organisms on which life-cycle or partial life-cycle toxicity tests have been conducted to determine an apparent or near no-effect response. Since the acute toxicity response of an organism varies among chemicals, and the toxicity of a chemical varies among organisms, it would be ideal if these values had some kind of constant factor, whether related by chemical or by species.

Experimental MATC values on 31 chemicals for fish taken from the literature are summarized in Table 1. By inspection it is obvious that they have not been determined to a precise concentration, only to within a range of concentrations, often within a 2-fold difference, but sometimes 5-fold. The accuracy of the MATC value for a given species can be no better than that determined by the closeness of the concentrations bracketing the precise MATC. In these statistical calculations, MATC values for each species and compound were averaged.

From the literature data shown in Table 1, it is apparent that MATC values can vary extensively among chemicals for the same species (>1,000 fold) and among species for the same chemical (>100 fold). MATC values are not useful for predicting other MATC values for unrelated species and unrelated chemicals. Where experimental MATC's are not available, there is a need for a time- and money-saving method of estimating or determining such data.

TABLE 1.—*Application factors for fish calculated from acute LC50 and MATC data.*

Chemical	Species	LC50[a] or LTC[a,b] (ppb) (4 days)	Range of MATC's (ppb)	Type of chronic life-cycle test C=complete P=partial	Range of application factors: MATC/LC50 or MATC/LTC	Reference[c]
\multicolumn{7}{c}{*Chlorinated hydrocarbon insecticides and PCB's*}
Lindane	Fathead minnow	69[d]	9.1–23.5	C	0.13–0.34	13
	Bluegill	30[d]	9.1–12.5	C	0.30–0.42	13
	Brook trout	26[d]	8.8–16.6	P	0.34–0.64	13
Chlordane	Fathead minnow	36.9	0.32	C	0.009	5
	Bluegill	59	0.54–1.2	P	0.009–0.02	5
	Brook trout	47	0.32–0.66	P	0.007–0.014	5
	Sheepshead minnow	12.5	0.5–0.8	C	0.04–0.06	24
Heptachlor	Fathead minnow	7[d]	0.86–1.84	C	0.12–0.26	15
	Sheepshead minnow	10.5	0.97–1.9	P	0.09–0.18	9
Endrin	Sheepshead minnow	0.38	0.12–0.31	C	0.35–0.91	9
Pentachlorophenol	Sheepshead minnow	442	47–88	C	0.11–0.20	24
Endosulfan	Fathead minnow	0.86[d]	0.2–0.4	P	0.23–0.47	15
Toxaphene	Fathead minnow	5.3[b]	0.025–0.054	P	0.005–0.01	18
	Channel catfish	15[b]	0.129–0.299	P	0.009–0.020	18
	Brook trout	4.1[b]	<0.039	P	<0.010	17
DDT	Fathead minnow	48	0.09	C	0.019	12
Methoxychlor	Fathead minnow	8.63	<0.125	P	<0.014	20
	Yellow perch	22.2	<0.625	P	<0.028	20
	Sheepshead minnow	49	12–23	P	0.24–0.47	9, 25
\multicolumn{7}{c}{*Phosphate and carbamate insecticides*}
Malathion	Fathead minnow	10,750[b]	200–580	C	0.019–0.053	7, 21
	Fathead minnow	9,000(S)	200–580	P	0.022–0.064	7, 21
	Bluegill	82[b]	3.6–7.4	P	0.043–0.090	7
	Bluegill	108	3.6–7.4	P	0.033–0.069	7
	Flagfish	349	8.6–10.9	C	0.019–0.054	10
	Sheepshead minnow	51	4.0–9	P	0.08–0.18	9, 25
Fenthion	Fathead minnow			C	0.07–0.12	16
Diazinon	Fathead minnow	7,800(S)	6.8–13.5	P	0.0009–0.0017	2
	Brook trout	770	<0.8–2.4	P	0.001–0.003	2
Azinphosmethyl	Fathead minnow	1,900 (S)	0.51–1.8	C	0.00009–0.00027	1
Carbaryl	Fathead minnow	9	0.21–0.68	C	0.023–0.075	6
Carbofuran	Sheepshead minnow	386	15–23	C	0.04–0.06	9, 24
\multicolumn{7}{c}{*Miscellaneous organic chemicals*}
LAS (linear alkyl benzene sulfonate)	Fathead minnow	4,350	630–1,200	P	0.14–0.28	28
2,4-D(BE)	Fathead minnow	5,600 (S)	300–1,500	P	0.054–0.268	21

TABLE 1.—*Continued.*

Chemical	Species	LC50[a] or LTC[a,b] (ppb) (4 days)	Range of MATC's (ppb)	Type of chronic life-cycle test C=complete P=partial	Range of application factors: MATC/LC50 or MATC/LTC[b]	Reference[c]
Propachlor	Fathead minnow			C	0.17–0.36	16
Dinoseb	Lake trout	77 (S)	<0.5	P	<0.006	29
Trifluralin	Fathead minnow	115[d]	1.9–5.1	P	0.017–0.044	15
	Sheepshead minnow	190	1.3–4.8	C	0.007–0.025	24
Butralin	Fathead minnow			C	0.2–0.42	16
Picloram	Lake trout	2,700 (S)	<35	P	<0.012	29
Atrazine	Fathead minnow	15,000 (S)[d]	210–520	P	0.014–0.035	14
	Bluegill	6,700[d]	90–500	P	0.014–0.075	14
	Brook trout	4,900[d]	60–120	P	0.013–0.024	14
Acrolein	Fathead minnow	84[d]	11.4–41.7	P	0.14–0.50	15
Captan	Fathead minnow	64[b]	16.5–39.5	P	0.25–0.61	11
Aroclor 1242	Fathead minnow	15[e]	0.86–5.4	C	0.057–0.14	23
	Fathead minnow	234–300[f]			0.0028–0.018	
Aroclor 1254	Fathead minnow	7.7[e]	0.52	C	0.064	23
	Fathead minnow	>33[f]			<0.016	

Metals[g]

Chemical	Species	LC50 or LTC (ppb)	Range of MATC's (ppb)	Type of chronic test	Range of application factors	Reference
Copper (sulfate)	Fathead minnow	75	10.6–18.4	P	0.14–0.24[h]	22
	Fathead minnow	470	14.5–33.0	P	0.031–0.07[i]	22
	Bluegill	1,100	21–40	P	0.019–0.036	3
	Brook trout	100	9.5–17.4	P, C	0.095–0.17	19
Cadmium (sulfate)	Fathead minnow	7,200	37–57	P	0.0051–0.0079	27
	Bluegill	20,400	31–80	P	0.0015–0.0039	8
Nickel (chloride)	Fathead minnow	27,000	380–730	P	0.014–0.027	26
Zinc (sulfate)	Fathead minnow	9,200	32–180	P	0.003–0.02	4

[a] Tests are flowing-water, except (S) denotes static test.
[b] Mean lethal threshold concentration (LTC).
[c] References:
  (1) Adelman et al. 1976
  (2) Allison and Hermanutz 1977
  (3) Benoit 1975
  (4) Brungs 1969
  (5) Cardwell et al. 1977
  (6) Carlson 1971
  (7) Eaton 1970
  (8) Eaton 1974
  (9) Hansen and Parrish 1977
  (10) Hermanutz 1977
  (11) Hermanutz et al. 1973
  (12) Jarvinen et al. 1976
  (13) Macek et al. 1976a
  (14) Macek et al. 1976b
  (15) Macek et al. 1976c
  (16) Macek and Sleight 1977
  (17) Mayer et al. 1975
  (18) Mayer et al. 1977
  (19) McKim and Benoit 1971
  (20) Merna and Eisele 1973
  (21) Mount and Stephan 1967
  (22) Mount and Stephan 1969
  (23) Nebeker et al. 1974
  (24) Parrish et al. 1978
  (25) Parrish et al. 1977
  (26) Pickering 1974
  (27) Pickering and Gast 1972
  (28) Pickering and Thatcher 1970
  (29) Woodward 1976
[d] Incipient LC50 used.
[e] Newly hatched fry.
[f] Two-month-old fish.
[g] LC50's are on a metal ion basis.
[h] Soft water.
[i] Hard water.

Applicaton Factors

Mount and Stephan (1967) proposed the use of an applicaton factor (AF) based on need to predict chronic fish toxicity (MATC) from acute toxicity data on the same species and chemical. The AF is presently obtained by dividing the 96-hour LC50 (or incipient LC50) into the MATC value. This calculation always results in a value less than one. In flow-through tests where concentrations occur which are too high to keep in solution, an "incipient" acute LC50 is calculated when aquatic organisms mortalities fall below 50%. This is done by calculating a linear regression equation from the known concentration-mortality data from concentrations which are soluble, as done by Macek and Sleight (1977).

AF data for fish and the LC50 and MATC values used for AF calculations are given in Table 1. The AF's for fish vary as much as 4–5 orders of magnitude (10,000–100,000 times) between some compounds. The limited fish data show that AF's *for a given compound* usually vary less than one order of magnitude between species. This variation is reasonable for predictive use of AF's in translating acute data to chronic toxicity data in other species of fish. There is no consistent pattern of MATC or AF values among all chemicals and species. AF data do not appear to be much more useful than MATC's for calculating MATC's between species except when extreme differences occur in LC50 values (such as with malathion). AF's do appear to be much more predictive between chemicals than MATC's.

In an effort to demonstrate a relationship between LC50 values and AF's, the fish data for both of these parameters and from all of the chemicals in Table 1 were classified into order-of-magnitude categories and their values plotted against each other. The random distribution of points indicates no predictable mathematical relationship within three orders of magnitude between the 48 comparative examples of the fish LC50 and AF values for these chemicals.

A major weakness in the calculation of the AF is its dependence on the use of the LC50 value. Since the LC50 is based on lethality, which is used to calculate the AF by use of an MATC which is based on unspecified *non*-lethal properties, the comparison of LC50 and MATC's is not based on equivalent or comparable measurements of toxicity. This is especially true when the MATC values take into account variables which include compound-specific no-effect levels such as enzymatic and reproductive effects. Also the 96-hour LC50 is determined at a time when many chemicals have reached a residue plateau level in small fish, but this short length of time for measuring effects obviously does not allow for delayed effects that may result from residue distribution to specific key tissues nor for chronic toxicity.

Applicaton Factor Equations

An application factor equation proposed here is meant to help bridge the difference between MATC's from different species (i.e., fish) on one compound. A good example is the variation in MATC of over 50-fold for malathion compared to a 4-fold variation in AF's (Table 1).

There is a need to determine whether the use of chronic no-effect data (MATC) is needed separately for every species of fish or whether predictive values for other species can be drawn from a determined value on one species by the use of equations derived from experimental data. If an equation could be used for each species it would save considerable apprehension by aquatic biologists and save much time and money spent by government, institutional, and industrial researchers to obtain experimental MATC's.

To test the validity of using AF's for this new concept, 40 fish AF's determined from 25 different compounds (see Table 1) were examined for uniformity and predictability by species. No attempt was made to correlate AF's among all compounds and all species since the data indicate this is obviously not very useful in a mixture of widely varying chemical structures.

Although the data base is sparse for comparison among bluegill, brook trout, and fathead minnow, the data show all correlation coefficients (r) of AF's of these species to be significant (see Table 2). Correlation coefficients between brook trout and fathead minnow for all chemicals are very good. Correlation between bluegill and fathead minnow and between brook trout and bluegill is only fair at the 5% level of significance. Equations for extrapolation of AF values from one of these species to another are given in Table 2. The AF equations offer a

TABLE 2.—*Interspecies equations for predicting bluegill (BG), brook trout (BT), and fathead minnow (FM) application factors (AF). OM = orders of magnitude.*

Fathead minnow AF to brook trout AF

$$\log(FM) = 0.25 + 1.15 \log(BT) \pm 0.22(OM)$$

$$N = 6; r = 0.996; P \leq 0.01$$

Brook trout AF to fathead minnow AF

$$\log(BT) = -0.23 + 0.86 \log(FM) \pm 0.21(OM)$$

$$N = 6; r = 0.996; P \leq 0.01$$

Bluegill AF to fathead minnow AF

$$\log(BG) = -0.29 + 0.82 \log(FM) \pm 1.42(OM)$$

$$N = 6; r = 0.76; P \sim 0.05$$

Bluegill AF to brook trout AF

$$\log(BG) = -0.52 + \log(BT) \pm 2.13(OM)$$

$$N = 4; r = 0.82; P \sim 0.05$$

---

method to bridge the relationships between chemicals as well as species. Assuming a meaningful correlation occurs, such AF equations can be developed for other species when the supporting data become available.

Comparison of Various Chronic Toxicity Indicators for Fish

The relative predictive usefulness of the previously discussed indicators of chronic toxicity (MATC's) is difficult to assess. One way to show differences between them is to analyze their respective use for predictive values by the breadth of the species and chemicals usefulness (i.e., one chemical—one species versus per mutations of related and unrelated species and related and unrelated chemicals).

The usefulness of the various chronicity indicators for predictions of unknown MATC values seems to be as follows:

(1) LC50—Not useful for calculations of MATC's for most other chemicals or species.

(2) LC50 × 0.1 application factor—Not useful for estimation of MATC's except as an "average." Experimentally derived AF's vary from about 0.1 to 0.001.

(3) MATC's—Not useful for estimation of MATC's between species of fish, but has limited use on one species.

(4) AF's—Useful on a given compound between species of fish where LC50 species sensitivity varies greatly. Such AF's, when divided by the LC50, can be used to estimate MATC's.

(5) AF equations—Useful for a limited number of chemicals to adjust for AF variations between species. Such AF's can then be used to estimate MATC's.

However, with various species of chemicals, the above uses of chronicity indicators are not uniformly applicable nor always best. The use of chronic toxicity data from other organisms and the use of physical and chemical properties of the compounds are often as useful or more useful in decision making than the various chronic toxicity indicators.

Use of Indicators of Chronic Toxicity for Daphnids

Available LC50, MATC, and AF data concerning *Daphnia magna* taken from the literature are presented in Table 3 for organic pesticides and in Table 4 for metals.

An advantage of using daphnids is their short life cycle and the opportunity to compare LC50 values for 48 hours and three weeks (a life cycle) as well as with reproduction data. The data of Biesinger and Christensen (1972) in Table 5 show the differences in chronic lethality and reproductive effects (i.e., effect on the number of instars), an important interpretive toxicological indicator. For example, tin did not show chronic lethality to the adult daphnid, but did strongly affect reproduction, in contrast to cobalt which showed chronic lethality, but not much additional effect on reproduction.

Predictive Use of Acute and Chronic Toxicity Data from Fish and Daphnids

Pearson et al. (1979) determined the acute LC50's on fathead minnow and *Daphnia magna*

TABLE 3.—*Application factors for* Daphnia magna *calculated from acute LC50 and MATC data from organic pesticides.*

Chemical	*Daphnia* LC50[a] (ppb)	MATC[b] (ppb)	Application factor MATC/LC50	Reference
Lindane	485	11–19[c]	0.02–0.03	Macek et al. 1976a
Chlordane	28.4	12.1–21.6[d]	0.33–0.76	Cardwell et al. 1977
Toxaphene	10	0.07–0.12	0.007–0.012	Mayer et al. 1977
Heptachlor	78	12.5–25.0[c]	0.16–0.3	Macek et al. 1976c
Endosulfan	166	2.7–7.0[c]	0.016–0.042	Macek et al. 1976c
Trifluralin	193	2.4–7.2[c]	0.012–0.037	Macek et al. 1976c
Acrolein	57	16.9–33.6[c]	0.30–0.59	Macek et al. 1976c
Atrazine	6,900	140–250[c]	0.02–0.04	Macek et al. 1976b

[a] Static-water test, 48-hour exposure.
[b] Flowing-water test.
[c] Based on 3-week exposure for each of 3 generations.
[d] Based on 4-week exposure for 1 generation.

for 30 related compounds, mostly nitrobenzenes, nitrotoluenes, nitrophenols, and derivatives. The average LC50 for fathead minnows was 18.7 ppm and for *Daphnia* 16.6 ppm. The LC50's on some individual compounds were a maximum of 4-fold greater for fathead minnows and 7-fold greater for *Daphnia*. In general, the toxicity of these compounds to fathead minnow and *Daphnia* were not greatly different, and not as great as differences shown elsewhere between species of fish.

Maki (1977) studied the relationship of the chronic toxicity values of fathead minnows (1-year study) to *Daphnia magna* (3-week study) for six surfactants in flowing water. The comparison of no observed effect concentrations (NOEC = MATC) of these organisms exhibited a strong correlation ($r = 0.98$). NOEC values for the six surfactants (of variable structures) for the fish and *Daphnia* comparisons fell within a 5-fold factor range for each compound.

Maki also assembled additional chronic toxicity (NOEC) data from the current literature for these two test species and for 27 test substances representing metals, polychlorinated biphenyl isomers, and pesticide formulations which were similarly compared. Correlation coefficients (r) for these chemical groups were 0.74, 0.93, and 0.47, respectively. A correlation analysis demonstrated a good overall coefficient of 0.79 for

TABLE 4.—*Application factors for* Daphnia magna[a] *calculated from acute LC50 and MATC data*[b] *for metals.*

Metal ion	*Daphnia* LC50 (ppb) 48-hour	*Daphnia* LC50 (ppb) 3-week	MATC[c] (ppb)	Application factor (MATC/48-hour LC50)
Sodium	1,820,000	1,480,000	680,000	0.37
Calcium	464,000	330,000	116,000	0.25
Magnesium	322,000	190,000	82,000	0.25
Potassium	166,000	97,000	53,000	0.32
Iron	9,600	5,900	4,380	0.47
Tin	55,000	42,000	350	0.006
Zinc	280	158	70	0.25
Nickel	1,120	130	30	0.027
Lead	450	300	30	0.067
Copper	60	44	22	0.37
Cobalt	1,620	21	10	0.006
Cadmium	65[d]	5	0.17	0.003

[a] Static-water test.
[b] Biesinger 1972.
[c] Criteria = <16% reproductive impairment.
[d] No food; all others with food.

TABLE 5.—*Comparison of application factors for fish, daphnids, and mammals.*

Chemical	Application factor[a]			
	Fish		Daphnia	Mammals[b]
Lindane	0.13–0.64	(3 sp.)	0.02–0.03	0.006–0.016
Chlordane	0.007–0.02	(2 sp.)	0.43–0.76	0.001–0.035
Heptachlor	0.12–0.26		0.16–0.32	0.001–0.006
Endosulfan	0.23–0.47		0.016–0.042	0.02–0.14
Toxaphene	0.005–0.02	(3 sp.)	0.007–0.012	0.004–0.03
Methoxychlor	< 0.014–0.28			0.001–0.002
Malathion	0.019–0.09	(2 sp.)		0.002–0.006
Diazinon	0.0009–0.055			0.0002–0.0015
Azinphosmethyl	0.00009–0.0003			0.008–0.01
Carbaryl	0.023–0.075			0.01–0.03
Dinoseb	<0.006			>0.086
Trifluralin	0.017–0.044		0.012–0.037	0.01(?)
Picloram	<0.012			0.018
Atrazine	0.01–0.07	(3 sp.)	0.02–0.04	<0.0016
Acrolein	0.14–0.50		0.30–0.59	
Captan	0.25–0.61			0.00001
Copper (sulfate)	0.02–0.24	(3 sp.)	0.37	
Nickel (chloride)	0.014–0.027		0.027	
Zinc (sulfate)	0.003–0.02		0.25	
Cadmium (sulfate)	0.0015–0.008	(2 sp.)	0.003	

[a] Application factor = MATC (maximum acceptable toxic concentration) divided by acute LC50, essentially the no-effect level for mammals.
[b] Kenaga (1978) calculated from acute oral LD50's and no-effect levels for rats.

these data. Although significant outliers were evident from this correlation, these differences can be explained primarily by differences in modes of toxic action of these test substances.

Maki suggests that the relatively short life cycle of the test organism, 21-day duration of the test, use of small water volumes, ease in handling and high fecundity of the organism, and good correlation of 21-day chronic data with chronic fish toxicity data make *Daphnia* chronic tests an attractive alternative to the conduct of longer-term fish tests.

Comparison of Application Factors from Fish, Daphnids, and Mammals

There is a large body of acute and chronic toxicity data on chemicals for mammals, a smaller body of data on fish, and less still for daphnids. It is reasonable to make use of such data through calculated application factors if such factors correlate well with each other. On the basis of data from the 20 chemicals shown in Table 5, a correlation of AF data was made among three groups of animal organisms. No significant correlation was shown between *Daphnia* and fish, fish and mammals, or mammals and *Daphnia* ($r = < 0.4$).

Conclusions

Over the past few years data have become available to help predict in a more orderly fashion the chronic toxicity of chemicals to fish and aquatic invertebrates. Certain measurements have been used to predict no-effect levels of chemicals. Some of their uses and limitations are as follows:

(1) The acute LC50 is highly variable between and within fish species and *Daphnia* for the various chemicals. The LC50 is not useful for predictions of chronic levels (MATC's) except for setting an upper limit (i.e., lower than the acute toxicity) on the given chemical and species tested. By the use of equations, LC50's of other species can be calculated for use on the same chemical.

(2) Maximum acceptable toxicant concentrations (MATC's essentially no-effect levels) are needed, especially for chemically sensitive aquatic organisms, to help set water quality standards. These MATC's can be obtained experimentally species by species, expensively and slowly. MATC's for other species can be calculated from experimentally derived MATC's, but are limited in accuracy to closely related species for the same compound. MATC's experimentally derived from critical life stages (usually

egg–fry) of fish appear to be good substitutes for MATC's derived from complete life-cycle toxicity tests. *Daphnia* tests offer the opportunity for quickly derived experimental MATC's, from reproduction tests, and LC50's as well as chronic lethality data. This combination of measurements helps detect the critically sensitive part of the life cycle. *Daphnia* MATC's correlate quite well with fathead minnow MATC's on a variety of chemical structures ($r = 0.76$) and even better with closely related structures ($r = 0.98$).

(3) Application factors (AF's) are used in the absence of an experimentally derived MATC to predict MATC's for other species of fish. Under some circumstances experimentally derived MATC's have value in predicting MATC's for related chemicals. When enough experimental data are available ($r = 0.75$), AF's can be predicted by the use of equations for unrelated chemicals and the specified organisms, as well as for specified chemicals and unrelated fish organisms, but probably not for a combination of both. AF's for fish, daphnids, and mammals did not correlate well.

It appears that, except for closely related organisms or chemicals the MATC's should preferably be derived experimentally using *Daphnia* life cycles or fish critical life stages, partial life-cycle tests rather than calculated from AF's.

Acknowledgments

I wish to thank Colin Park of The Dow Chemical Company for the statistical correlation of data and derivation of the equations presented in this paper.

References

ADELMAN, I. R., L. L. SMITH, JR., AND G. D. SIESENNOP. 1976. Chronic toxicity of guthion to the fathead minnow (*Pimephales promelas* Rafinesque). Bull. Environ. Contam. Toxicol. 15:726–33.

ALLISON, D. T., AND R. O. HERMANUTZ. 1977. Toxicity of diazinon to brook trout and fathead minnow. U.S. Environ. Prot. Agency Ecol. Res. Ser. EPA-600/3-77-060.

APHA (AMERICAN PUBLIC HEALTH ASSOCIATION), AMERICAN WATER WORKS ASSOCIATION, AND WATER POLLUTION CONTROL FEDERATION. 1975. Standard methods for the examination of water and wastewater, 14th ed. Am. Publ. Health Assoc., Washington, D.C. 874 pp.

ANDREW, R. W., D. A. BENOIT, J. G. EATON, J. M. MCKIM, AND C. E. STEPHAN. 1978. Evaluation of an application factor hypothesis. U.S. Environ. Prot. Agency, Environ. Res. Lab., Duluth, Minn, 51 pp. (Prepublication copy.)

BENOIT, D. A. 1975. Chronic effects of copper on survival, growth, and reproduction of the bluegill (*Lepomis macrochirus*). Trans. Am. Fish. Soc. 104:353–358.

BIESINGER, K. E., AND G. M. CHRISTENSEN. 1972. Effects of various metals on survival, growth, reproduction, and metabolism of *Daphnia magna*. J. Fish. Res. Board Can. 29:1691–1700.

BRUNGS, W. A. 1969. Chronic toxicity of zinc to the fathead minnow, *Pimephales promelas* Rafinesque. Trans. Am. Fish. Soc. 98:272–279.

CARDWELL, R. D., D. G. FOREMAN, T. R. PAYNE, AND D. J. WILBUR. 1977. Acute and chronic toxicity of chlordane to fish and invertebrates. U.S. Environ. Prot. Agency, Ecol. Res. Ser. EPA-600/3-77-019.

CARLSON, A. R. 1971. Effects of long-term exposure of carbaryl (sevin) on survival, growth, and reproduction of the fathead minnow (*Pimephales promelas*). J. Fish. Res. Board Can. 29:383–387.

EATON, J. G. 1970. Chronic malathion toxicity to the bluegill (*Lepomis macrochirus* Rafinesque). Water Res. 4:673–684.

EATON, J. G. 1974. Chronic cadmium toxicity to the bluegill (*Lepomis macrochirus* Rafinesque). Trans. Am. Fish. Soc. 103:729–735.

HANSEN, D. J., AND P. R. PARRISH. 1977. Suitability of sheepshead minnows (*Cyprinodon variegatus*) for life-cycle toxicity tests. Pages 117–126 *in* F. L. Mayer and J. L. Hamelink, eds. Aquatic toxicology and hazard evaluation. ASTM STP 634. Am. Soc. Test. Mater., Philadelphia.

HERMANUTZ, R. O. 1977. Endrin and malathion toxicity to flagfish (*Jordanella floridae*). Arch. Environ. Contam. Toxicol.

HERMANUTZ, R. O., L. H. MUELLER, AND K. D. KEMPFERT. 1973. Captan toxicity to fathead minnows (*Pimephales promelas*), bluegills (*Lepomis macrochirus*), and brook trout (*Salvelinus fontinalis*). J. Fish Res. Board Can. 30:1811–1817.

JARVINEN, A. W., M. J. HOFFMAN, AND T. W. THORSLUND. 1976. Toxicity of DDT food and water exposure to fathead minnow. U.S. Environ. Prot. Agency Ecol. Res. Ser. EPA-600/3-76-114.

KENAGA, E. E. 1978. Test organisms and methods useful for early assessment of acute toxicity of chemicals. Dow Chemical U.S.A., Midland, Mich. 28 pp. Manuscript submitted to Environ. Sci. Technol.

KENAGA, E. E. In press. Acute and chronic toxicity of 75 pesticides to various animal species. Down to Earth.

MACEK, K. J., K. S. BUXTON, S. K. DERR, J. W. DEAN, AND S. SAUTER. 1976a. Chronic toxicity of lindane to selected aquatic invertebrates and fishes. U.S. Environ. Prot. Agency Ecol. Res. Ser. EPA-600/3-76-046.

MACEK, K. J., K. S. BUXTON, S. SAUTER, S. GNILKA, AND J. W. DEAN. 1976b. Chronic toxicity of atrazine to selected aquatic invertebrates and fishes. U.S. Environ. Prot. Agency Ecol. Res. Ser. EPA-600/3-76-047.

MACEK, K. J., M. A. LINDBERG, S. SAUTER, K. S. BUXTON, AND P. A. COASTA. 1976c. Toxicity of four pesticides to water fleas and fathead minnows. Acute and chronic toxicity to acrolein, heptachlor, endosulfan, and trifluralin to the water flea (*Daphnia magna*) and the fathead minnow (*Pimephales promelas*). U.S. Environ. Prot. Agency Ecol. Res. Ser. EPA-600/3-76-099.

MACEK, K. J., AND B. H. SLEIGHT, III. 1977. Utility of toxicity tests with embryos and fry of fish in evaluating hazards associated with the chronic toxicity of chemicals to fishes. Pages 137–146 in F. L. Mayer and J. L. Hamelink, eds. Aquatic toxicology and hazard evaluation. ASTM STP 634. Am. Soc. Test. Mater., Philadelphia.

MAKI, A. W. 1977. Correlations between fathead minnow, *Pimephales promelas*, and *Daphnia magna* chronic toxicity values for several classes of test substances. Presented at 8th Conf. Use Environ. Toxicol., Univ. Calif., Irvine, Oct. 4–6. 30 pp.

MAYER, F. L., JR., P. M. MEHRLE, JR., AND W. P. DWYER. 1975. Toxaphene effects on reproduction, growth, and mortality of brook trout. U.S. Environ. Prot. Agency Ecol. Res. Ser. EPA-600/3-75-013.

MAYER, F. L., JR., P. M. MEHRLE, JR., AND W. P. DWYER. 1977. Toxaphene: chronic toxicity to fathead minnows and channel catfish. U.S. Environ. Prot. Agency. Ecol. Res. Ser. EPA-600/3-77-069.

McKIM, J. M. 1977. Evaluation of tests with early life stages of fish for predicting long-term toxicity. J. Fish. Res. Board Can. 34:1148–1154.

McKIM, J. M., AND D. A. BENOIT. 1971. Effects on long-term exposures to copper on survival, growth, and reproduction of brook trout (*Salvalinus fontinalis*). J. Fish. Res. Roard. Can. 28:655–662.

MERNA, J. W., AND P. J. EISELE. 1973. The effects of methoxychlor on aquatic biota. U.S. Environ. Prot. Agency EPA-R3-73-046. 59 pp.

MOUNT, D. I. 1977. An assessment of application factors in aquatic toxicology. Pages 183–190 in Recent advances in fish toxicology: a symposium. U.S. Environ. Prot. Agency EPA-600/3-77-085.

MOUNT, D. I., AND C. E. STEPHAN. 1967. A method for establishing acceptable toxicant limits for fish—malathion and the butoxyethanol ester of 2,4-D. Trans. Am. Fish. Soc. 96:185–193.

MOUNT, D. I., AND C. E. STEPHAN. 1969. Chronic toxicity of copper to the fathead minnow (*Pimephales promelas*) in soft water. J. Fish. Res. Board Can. 26:2449–2457.

NEBEKER, A. V., F. A. PUGLISI, AND D. L. DEFOE. 1974. Effect of polycholorinated biphenyl compounds on survival and reproduction of the fathead minnow and flagfish. Trans. Am. Fish. Soc. 103:562–568.

PARRISH, P. R., E. E. DYAR, J. M. ENOS, AND W. B. WILSON. 1978. Chronic toxicity of chlordane, trifluralin, and pentachlorophenol to sheepshead minnows (*Cyprinodon variegatus*). U.S. Environ. Prot. Agency, Gulf Breeze, Fla., EPA-600/3-78-010.

PARRISH, P. R., E. E. DYAR, M. A. LINDBERG, C. M. SHANIKA, AND J. M. ENOS. 1977. Chronic toxicity of methoxychlor, malathion, and carbofuran to sheepshead minnows (*Cyprinodon variegatus*). U.S. Environ. Prot. Agency, Gulf Breeze, Fla., EPA-600/3-77-059.

PEARSON, J. G., J. P. GLENNON, J. J. BARKLEY, AND J. W. HIGHFILL. 1979. An approach to the toxicological evaluation of a complex industrial wastewater. In L. L. Marking and R. A. Kimerle, eds. Aquatic toxicology. ASTM STP 667. Am. Soc. Test. Mater., Philadelphia.

PICKERING, Q. H. 1974. Chronic toxicity of nickel to the fathead minnow. J. Water Pollut. Control. Fed. 46:760–765.

PICKERING, Q. H., AND M. H. GAST. 1972. Acute and chronic toxicity of cadmium to the fathead minnow (*Pimephales promelas*). J. Fish. Res. Board Can. 29:1099–1106.

PICKERING, Q. H., AND T. O. THATCHER. 1970. The chronic toxicity of linear alkylate sulfonate (LAS) to *Pimephales promelas* Rafinesque. J. Water Pollut. Control Fed. 42:243–254.

WOODWARD, D. F. 1976. Toxicity of the herbicides dinoseb and picloram to cutthroat (*Salmo clarki*) and lake trout (*Salvelinus namaycush*). J. Fish. Res. Board Can. 33:1671–1676.

Adequacy of Laboratory Data for Protecting Aquatic Communities

Donald I. Mount

Environmental Research Laboratory—Duluth
US Environmental Protection Agency
6201 Congdon Boulevard, Duluth, Minnesota 55804

Abstract

When the protection of aquatic communities is discussed relative to water-pollution control efforts, we must recognize that adequacy of protection may be judged on a different basis by the public and by the ecologist: by the former on community features; by the latter on community functions. Public attitudes are important, however, because public support is crucial to the funding of both research and regulation. Community functions are often not perceived by the public as important.

In this paper I will first analyze the criteria by which the general public is likely to judge the protection of aquatic communities. Let the record be clear that such an analysis carries no implication as to the accuracy of such judgments. We cannot, however, belittle the importance of the public's judgment because it is the people who pay for research, pollution control, and waste treatment. I will also discuss some specific technical considerations that, in part, determine the adequacy of laboratory data as they are usually applied in practical situations.

We must recognize at the outset that, although we speak of aquatic communities as entities that have specific physical, chemical, and biological characteristics, such statements are gross oversimplifications. To a large extent a community is what the observer perceives it to be. Of course, the lines of demarcation are rather distinct at the gross level of definition. For example, the community at the edge of the pond will have primary producers, grazers, decomposers, and predators. But just as cities have city limits, manufacturers, buyers, garbage collectors, and shrewd retailers, the real "flavor" of a community (city) cannot be described by such gross characterization. Furthermore, a community (city) is not necessarily stable if only such characteristics are maintained. Clearly, more refined descriptions must be developed.

Thus the real problem emerges as we attempt to reveal the more intricate workings of the aquatic community. We find that it is composed of subcommunities, each with slightly different mixes of functions and species performing those functions. For example, the shore zone may have relatively fewer decomposers because wave action moves the organic matter elsewhere leaving little to be decomposed, but primary production may be relatively more significant because adequate sunlight reaches to the bottom in sufficient intensity to support photosynthesis. But where does the next subcommunity begin in which decomposition is more dominant than primary production because wave action has diminished, sedimentation is more prominent, and insufficient sunlight penetrates to support photosynthesis? Is the boundary the same at the surface and near the bottom?

We soon recognize that our nice, neat pond community is really a host of subcommunities, each grading into another, similar in some ways, yet quite different in others. The problem is even more perplexing when several pond communities are compared. Although each community performs many identical functions, the rates of performance are different and the species associations responsible for various functions are different. But how different must they be before we have a "different community"?

Predicting Community Effects

At the risk of insulting the reader's intelligence, I have perhaps belabored some rather simple attributes of aquatic communities to make the point that to decide if laboratory data adequately protect communities, we must be able to define the community precisely. When we try to precisely define the community, we find it is a group of subcommunities, none of which have very clear boundaries functionally or physically. So by what yardstick or units of measure do we judge community protection? What are the error bars upon which probability of protection can be based? How much change is "more than normal"?

Indeed, is a community a real entity? Or is it the end result of random organism interactions that are predictable, analogous to the gas laws that are precise but whose prediction depends on totally random behavior of molecules? Do we judge a community by the rate of community functions, or by species composition, or by both? Is a community dependent on man's influence a "natural community" or any less valuable than a "natural" one? Is a cold-water community (including trout) below a reservoir on a warm-water stream a natural community? An acceptable one? Is a warm-water fish community below a power-plant discharge on a cold-water stream a natural community? An acceptable one? Does the introduction of coho salmon in Lake Michigan allow adequate protection of the former community? Obviously, except in the most gross cases, adequacy of protection is a property only in the eyes of the beholder, i.e., an opinion! Figure 1 is my attempt to portray the "state of the art" of predicting community effects. Unquestionably, laboratory data can predict well if we are concerned with gross effects; and unquestionably, laboratory data cannot predict well if we are concerned with refined effects. Be aware that as much of the problem may lie with our ability to make "refined" descriptions of the community as with the validity of the laboratory data. So adequacy of protection is not only subjective but also a matter of degree.

Community Change versus Pollutant Concentration

A second important point is illustrated in Figure 2, namely that until the predicted change is great enough to exceed the normal community

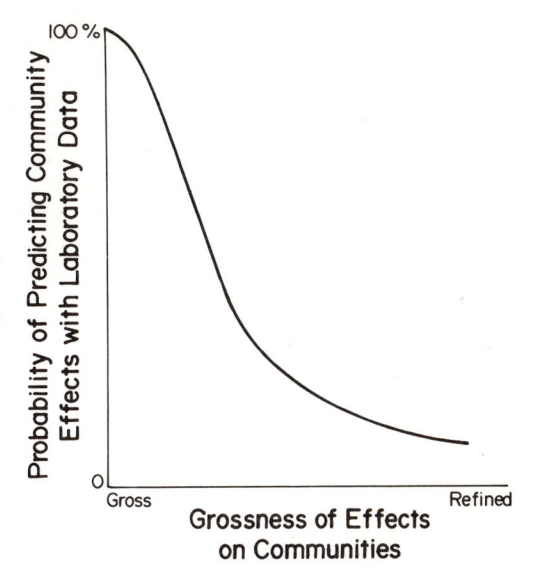

FIGURE 1.—*Probability of predicting community effects from laboratory data.*

variations, we cannot test the adequacy of laboratory data. Normal variation for some communities may even include extinction, leaving little place for judging adequacy of protection. Clearly then, the predictive capability of laboratory data depends on inherent characteristics of the community as well. The greater the normal community fluctuation, the less precise the prediction can be.

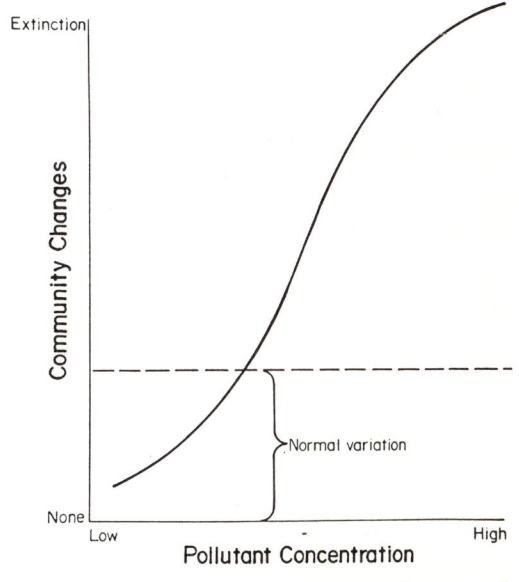

FIGURE 2.—*Relationship of community changes and pollutant concentrations.*

Community Tolerance

Figure 3 portrays community tolerance in relation to alternate pathways of community function. I have used tolerance rather than stability to emphasize the applicability to our concern at this meeting, that is, stress from pollutants. The foundation for the curve is simply a probability function—the more alternate pathways by which functions can proceed, the more will survive stress and be available for use. There is no implication here that particular species will not disappear, but only that their functions will be continued. So, biomass production from a community dominated by blue-green algae, sludge worms, and carp may be comparable to (from a community point of view) biomass production from diatoms, stoneflies, and trout. If we do not exclude outside inputs for community stabilization—man, for example—the carp community can be as stable as or more stable than the trout community. The discharge of sewage is rather constant and dependable enough to support a stable community. The pressure on the trout population from fishing may well be spotty and sporadic.

The point is, we have placed all sorts of phony and bizarre constraints on what is natural and unnatural, what is desirable and undesirable, what is change and what is not. As a result, judging the adequacy of protection becomes almost an individual matter. The confusion and lack of agreement that exists among scientists, industrialists, purists, environmentalists, regulators, and the unconcerned public are the factors chiefly responsible for much of our dilemma. A common perception prevails that the introduction of a species into a community, for example steelhead into Lake Superior or coho salmon into Lake Michigan, is a beneficial change that does not harm "biological integrity," improves the community, and makes it more useful to society. At least that is how the fishermen view the introduction and, by and large, public attitude seems to be supportive. Population dynamics and community function dictate that one or more other species will decline as a result of this introduction. Another example: a trout fishery below a reservoir on a warm-water stream and in the cool hypolimnetic water is valued immensely by the public even though the warm-water fishery is destroyed.

On the contrary, the discharge of heat from a power plant that results in more catfish and fewer perch is usually considered unacceptable, adverse, damaging, and destructive of the balanced indigenous population. Great sums of capital are likely to be demanded to avoid such a change.

From the community-function viewpoint, there is no basis to judge that one situation is different from another. All are likely to be stable, and certainly community functions such as carbon fixed, biomass produced, P/R ratios, decomposition rate, and dissolved oxygen evolution will all be within "normal range."

So by what yardstick is the increase in catfish from a power plant labeled unacceptable?

Social and Ecological Value of Community Function

I think it is obvious that the reason has to do with human psychology. First of all, the change has been caused by the waste products of man's activity whereas the introduction of steelhead and trout below the reservoir was deliberate and not the result of industry. Communities sustained by industrial waste products (carp, catfish) are a priori undesirable, unnatural, and unacceptable in the public view. Well-planned in-

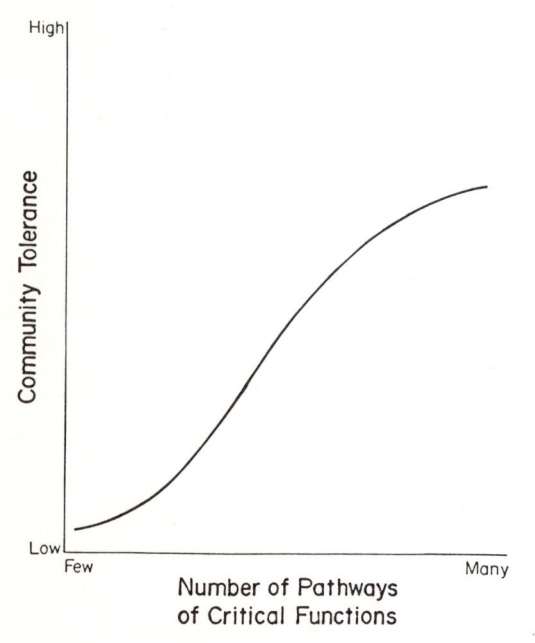

FIGURE 3.—*Relationship of community tolerance and number of functional pathways.*

troductions are acceptable unless they become pests—to wit, carp in Lake Erie. Second, the public tends to place more value on scarce or expensive species—snobbishness, if you will. Few objects that are easy to get, common, and cheap are highly valued. Indeed, much of today's fish management is directed at getting species to live where they are not best suited to live. The public will place highest value on the species most difficult to maintain—hardly an ecologically based principle!

The above considerations are summarized in Figure 4. Here we see that social value is highest for specific community functions such as biomass (and even size of units of biomass) of particular species such as steelhead or salmon. As the function becomes more generalized, let us say total biomass of fish produced, social interest wanes, because John Q. Public does not really care about the biomass of *Noturus flavus* or *Notropis atherinoides*. Finally, at the general end of the spectrum public concern may be nonexistent. The amount of carbon fixed is a function unknown to most of society, let alone of concern to them. On the other hand, the general end of the spectrum is of most vital concern to the maintenance of an aquatic community and to the ecologist. If carbon is not fixed, there will be no community, but on the contrary, communities can and do function well without any fish biomass and certainly without steelhead and salmon. Other generalized functions such as decomposition and nitrification are essential functions, the rates of which must be within certain bounds to keep the community "in tune." But do you know of one instance where decomposition rate has been seriously considered when water quality standards were set?

Relationship of Value and Community Size

Yet another generalization that has an impact on our judgment of community protection is shown in Figure 5. Here we see that the social value of the community has a bimodal relationship to the size or area of the community. For example, very small communities, such as the short stretch of cold water below the dam, or the shore of Lake Superior, have a very high value in the public eye. The whole of the Tennessee River or of Lake Superior, however, has relatively less perceived social value because we do not fish or swim in the water body and we cannot readily see all of it. Very large communities such as open oceans or tropical rain forests become very important to us socially and economically because we see (with coaching) that they are responsible for functions on a large scale that are necessary for the continued survival of man— that is, O_2-CO_2 balance of the atmosphere and cellulose production. We may only now recognize their vital functions as they become threatened by toxic organic substances, metals, and acid rain. Thus, the very large communities have more social value than small ones because our survival depends on them. Small communities only titillate our fancy. Large ones enable us to live. An appreciation of the above considerations is essential to the judgment of adequate community protection. You will perhaps see that the ecologically important community functions are not those that the public wishes to obtain from communities. Rather the public wants specific outputs that are not important community characteristics, that is, steelhead, not biomass. We do not even want catfish biomass if it is enhanced by waste discharges!

The regulation of water pollution is the principal vehicle for protecting aquatic communities.

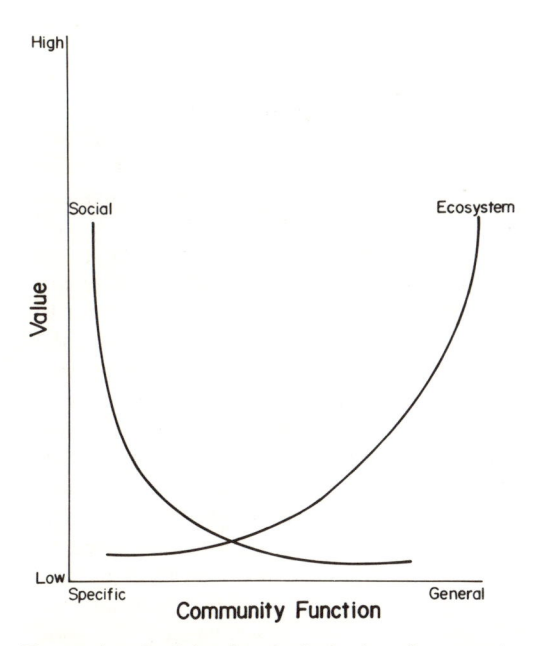

FIGURE 4.—*Social and ecological value of community function.*

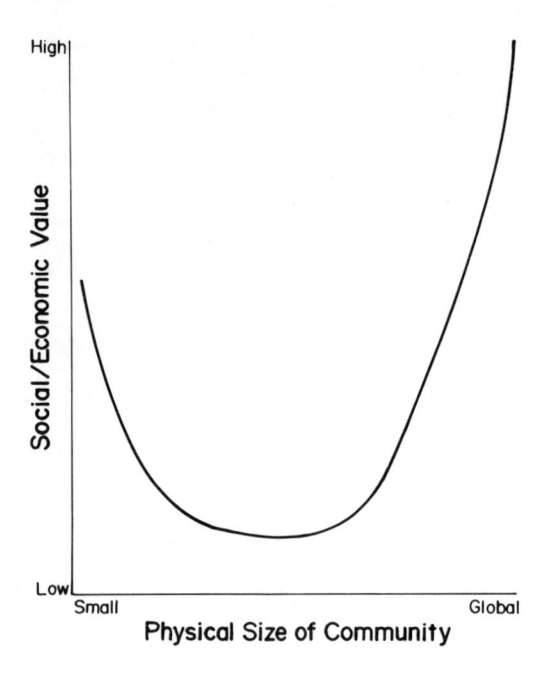

FIGURE 5.—*Relationship of value and community size.*

Whether scientists like it or not, the success of regulatory efforts will be judged by the cost of water, the cost of waste treatment (in dollars, in energy, and in convenience), the availability of swimming waters, aesthetics, and the quality of sport fishing, among other things. Therefore, the success of protecting aquatic communities will be judged in part by the number, size, and *catchability* of steelhead, trout, and selected other species, but not by the fish biomass, carbon fixed, P/R ratio, decomposition rate, or species diversity. In view of this situation, perhaps a more pertinent title for this paper would be "Are Laboratory Data Adequate for Protecting Fish Species of Sport or Commerical Value to Man?" This new title drives to the essence of how success will be judged!

In all probability as the flux of elements and compounds continues to increase as a result of man's activities, our pollution-control efforts will need to change from protecting steelhead in Lake Superior and sauger below Wheeler Reservoir to such matters as reducing acid rain and global fallout of mercury and protecting and using the assimilative capacity of communities. Just as Figure 5 depicts, we will ultimately see that community functions viewed from a global scale are socially as well as ecologically more important than small highly valued communities. If acid rain threatens our food supply, carp may assume an unprecedented glory!

Relevance of Laboratory Data

I would like now to turn to other practical reasons why laboratory data of the type currently being used for establishing standards may or may not be protective enough for communities.

Perhaps the point of most concern relative to the insufficient protection of laboratory data is that the sensitive species have not been tested. This concern may be more valid for marine than for freshwater systems because a smaller percentage of the marine species has been tested. In reviewing the criteria documents prepared by the Environmental Protection Agency for the 65 groups of chemicals involved in the Consent Decree, I was surprised to see that the difference in species sensitivity for given chemicals was so small. Since these documents collectively cover many chemicals and a cross section of species, one would expect to see some large and frequent species differences. Within these 65 groups of chemicals, species commonly differ in sensitivity by as much as an order of magnitude, but such differences are not large when viewed against the impact of flow and discharge variability, natural variation, and seasonal effects.

Another concern is that the organisms face stress, disease, and predation in aquatic communities, whereas test organisms in the laboratory are healthy, well fed, and disease-free. The argument proposed is that for these reasons laboratory organisms will be more tolerant to pollutants. Strangely enough, when "the other side" is attacking laboratory data in adversary cases, they use the very same line of evidence to establish that laboratory results are biased to the more toxic end of the spectrum, arguing that animals in test chambers are stressed, likely to be diseased, fed inadequate diets, and are likely to be injured in handling or by bumping the sides of the test chamber. The latter arguments may have been true some years ago before toxicity-testing techniques were as well developed as they now are. With advent of flow-through techniques, better knowledge of fish nutrition, antibiotics, and the use of acclimation and good water supplies, laboratory conditions probably are optimum for most organisms. The longevity of fish held in testing laboratories, our ability to consistently complete life-cycle tests, and the ability to hold large numbers of fish in crowded conditions, all suggest that laboratory conditions are generally quite suitable.

A third common concern is the presence of other toxic chemicals and of synergism. With regard to synergism, the evidence for concern is hard to find. Certainly Doudoroff (1952) and some of the workers in England have conclusively demonstrated that there may be synergism between some of the metals in acute high-level exposures. Other English workers, Lloyd and Jordan (1964), have found only additive effects when testing complex mixtures in lethal exposures.

The evidence for synergism is virtually non-existent in exposures of long duration and at concentrations that are near the no-effect levels. Biesinger, Christensen, and Fiandt (unpublished data 1973) found a mixture of 21 metals only slightly additive in its effect on *Daphnia*. Each metal, when present at a concentration that caused a 16% reproductive impairment individually, in combination reduced reproductive impairment only by about 80%. While not definitive, this work suggests that the components were less than strictly additive.

Eaton (1973) examined the chronic effect of zinc, copper, and cadmium in life-cycle exposures of fathead minnows and concluded that copper toxicity may have been slightly enhanced and cadmium toxicity slightly suppressed in the mixture of the metals.

Time does not permit me to review here our present knowledge regarding the effect of mixtures. The generally accepted concensus among toxicologists seems to be that, for the most part, synergism occurs mainly in high-level exposures and infrequently in concentrations near the no-effect level. Logically, there seems to be no a priori reason why antagonism should not be as common as synergism. If so, then on the average the effects may cancel each other.

Usually several conditions in toxicity tests will bias results toward over-estimating toxicity. Generally, we usually use constant exposure at constant concentrations. The use of constant concentrations is probably the less realistic of the two conditions because in very few situations are concentrations constant, but exposure may be constant. For pollutants not naturally found in water, continuous exposure is also likely to give results biased toward overestimation of toxicity. We usually use chemical forms most toxic in a water quality where the most toxicity will be manifested: for example, soft water and lead chloride. In most laboratories a water supply is used that is lower in suspended solids and organic matter than many surface waters. These frequently make chemicals less biologically active, and toxicity is overestimated. Especially in tests where reproduction is not measured, temperatures toward the upper end of the range are usually chosen. However, high levels of dissolved oxygen and neutral pH values are maintained, which would favor underestimation of toxicity. Finally, we usually do not allow acclimation to the toxicant, and therefore do not model the more common circumstance except for spills and other emergencies.

At least two characteristics of the application of test data in the standard setting favor some degree of overprotection. First, discharge levels often are based on 7-day, 10-year low flow, which ensures that for a very high percentage of the time concentrations will be less in the receiving water.

Second, no-effect values from chonic tests are usually applied "anytime-anyplace" at the edge of mixing zones, assuring that, for most of the time and area, concentrations will be less.

One significant difficulty in judging the validity of laboratory-based predictions of field effects is the inadequate data base for determining the exposure conditions in the field. Only rarely do we have daily measurements from a group of well-placed sampling stations so that the exposure history can be ascertained. Therefore, we really cannot assess the validity of predictions in many field situations and that is why the stream study by Geckler et al. (1976) is so valuable.

In the Geckler study copper was introduced proportional to flow into a natural stream for about 3 years. The copper produced an effect at a concentration about 50% of the one predicted from laboratory data to have no effect. The effect was behavioral avoidance, which had not been measured in laboratory tests, but was very pronounced in the stream and resulted in a mass migration of animals from the test area. Geckler and his coworkers failed to find indirect food-chain effects, but they did show that a copper concentration that had little or no effect on reproduction during the summer spawning season when pH and alkalinity were high was lethal in a few days during spring runoff when pH and hardness were lower. This study provides the best insight into how well data pertaining to maximum available toxiciant concentrations (MATC's) predict community effects but not functional ones, because exposure conditions are better described. If we are indeed able to predict within an error factor of 2×, then we probably have predictive capability exceeding our ability to achieve desired conditions through waste treatment and controlled discharges.

A recently completed study by Wrenn and Forsythe (1978) on the effects of temperature on walleye demonstrated that laboratory data overestimated thermal effects. These authors suggest that the acclimation regime and direct transfer techniques were probably responsible for the discrepancy. In any event, the differences between laboratory and field data in these two studies are certainly not great.

Conclusions

In conclusion I would like to refer to the earlier part of this paper and emphasize that protection of the aquatic community, as the ecologist views it, is probably not the goal of most of the populace. People seem to desire the protection of a few selected species especially pleasing to them. Community functions such as biomass, carbon fixed, and decomposition are not important as the public perceives the situation. Emerging global problems such as acid rain may change that perception. I see no way to balance the ledger objectively in view of all the above considerations and place confidence intervals on the ability to protect aquatic communities by using laboratory data based on single species. It is clear we have made a quantum leap forward in the past quarter century, and our predictions are certainly far better now than when we only had static tests of 96 hours' duration. We have much to do yet to improve our accuracy and precision, but in my opinion the prospects are not at all as hopeless as some would have us believe. Instances where communities were not protected when state-of-the-art laboratory tests were used are not easy to cite.

References

BIESINGER, K. E., G. M. CHRISTENSEN, AND J. T. FIANDT. 1973. The effects of metal mixtures on *Daphnia* reproduction. Nat. Water Qual. Lab. Interim Rep. on Work Planned in ROAP 16AAD, FY 73, Task 11-A:2-7.

DOUDOROFF, P. 1952. Some recent developments in the study of toxic industrial wastes. Pages 21–25 *in* Proc. 4th Annu. Ind. Waste Conf., March 26, 27, 1952. State Coll. Wash., Pullman.

EATON, J. G. 1973. Chronic toxicity of a copper, cadmium and zinc mixture to the fathead minnow (*Pimephales promelas* Rafinesque). Water Res. 7:1723–1736.

GECKLER, J. R., W. B. HORNING, T. M. NEIHEISEL, Q. H. PICKERING, AND E. L. ROBINSON. 1976. Validity of laboratory tests for predicting copper toxicity in streams. US Environ. Prot. Agency Ecol. Res. Ser. 600/3-76-116.

LLOYD, R., AND D. H. M. JORDAN. 1964. Predicted and observed toxicities of several sewage effluents to rainbow trout: a further study. J. Proc. Inst. Sewage Purif. Part 2: 3–6 pp.

WRENN, W. B., AND T. D. FORSYTHE. 1978. Effects of temperature on production and yield of juvenile walleyes in experimental ecosystems. Am. Fish. Spec. Publ. 11:66–73.

Discussion Synopsis—
Hazard Assessment Philosophy and Principles

R. A. KIMERLE, *Chairman*

DEAN BRANSON, GEORGE BAUGHMAN, KARL BESCH, GIL VEITH, KARIM AHMED, CURT HUTCHINSON, RICHARD PURDY, PETER HODSON, AND WILLIAM BISHOP

At the 1977 Pellston Workshop on Estimating the Hazard of Chemical Substances to Aquatic Life and in the proceedings of that workshop (Cairns et al. 1978), a number of consensus principles emerged which represented the state of the art of hazard evaluation of aquatic life. The discussion initiation papers and discussion at the Waterville Valley Workshop session on Hazard Assessment Philosophy and Principles presented an opportunity to review and further analyze those principles and to add new ideas. Basic to assessing the hazards of chemical substances to aquatic life is the principle that an evaluation of hazard can be obtained by a comparison of the exposure concentration to the effects concentration. In general, participants in this workshop felt that the basic philosophy of arriving at a hazard assessment was appropriate. It was also recognized that some sort of decision criteria was appropriate for helping to decide when to (1) discontinue consideration of use of a chemical because of unacceptable risk, (2) recommend use of a chemical because of acceptable risk, and (3) continue the testing program because of marginally acceptable risk.

The discussion initiation papers presented in the preceding session—Hazard Assessment Philosophy and Principles—added to our knowledge of how to assess the hazards of chemicals to aquatic life by introducing a couple of new principles and by giving us a better ability to interpret and use our data base. Specific conclusions of this session are numbered below.

(1) A major concept of the earlier Pellston Workshop (Cairns et al. 1978) was that environmental risk of a chemical could be assessed by examining the following ratio:

$$\frac{\text{No-Effects Concentration (NEC)}}{\text{Estimated Environmental Concentration (EEC)}}$$

As more data are generated in a sequential testing program, the confidence intervals around the no-effects concentration and the estimated environmental concentration became smaller. Figure 1, taken from Maki's paper in this Session, illustrates this concept. The participants of the workshop identified several deficiencies in the way this principle is presented. There are two principal deficiencies of Figure 1.

(a) *The estimated environmental concentration (EEC) and no-effects concentration (NEC) appear constant in Figure 1*. In reality this is not true. Varying EEC's and NEC's result in a fluctuation risk. If estimates of these variations are available, the probability of occurence of an extreme high EEC and an extreme low NEC can be calculated. The result is a refinement of risk estimation and an ability to estimate the frequency that a given risk level is exceeded during any fixed time period (e.g., a summer season in which flow is low and temperature high).

(b) *The EEC is not a fixed level but can be manipulated by "risk management" practices*. A ratio of NEC of EEC approaching unity indicates a high risk of adverse effects. Rather than not producing a chemical, a company may desire to restrict use, restrict use pattern, treat wastes, etc., to reduce chemical release to the environment and hence the EEC. This may be called "risk management." In the extreme, a "contained use" may be specified. However, this term should not be used since it implies no environmental release. Since no system is perfect, there is a given probability of loss that is a function of the safety precautions. Consequently, a

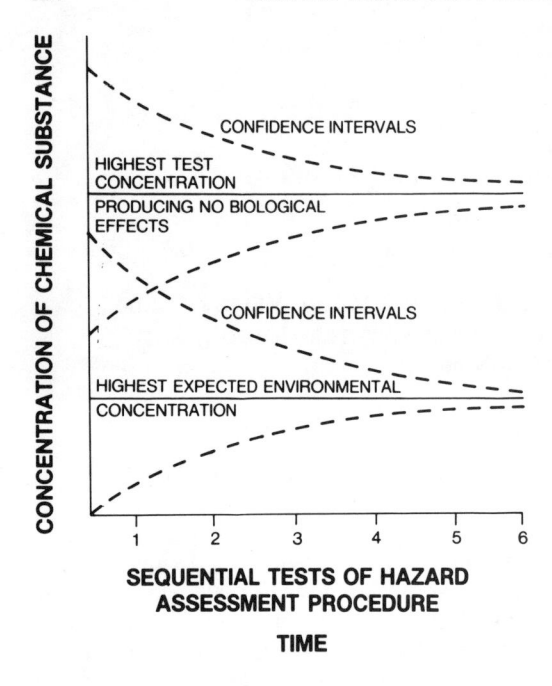

FIGURE 1.—*Relationship of no-effect concentration and estimated environmental concentration of a chemical, and their confidence limits, through a sequential testing program.*

desired EEC (and risk) could be specified to back-calculate the required probability of environmental contamination and the safety procedures to meet that probability.

(2) Early tier hazard assessment is not only useful for decision making regarding the extent of additional environmental research required for new chemicals, but could also be useful to identify those chemicals that should be included or excluded from a list of chemicals needing establishment of water quality standards. The data from tests used in the data base of hazard assessment schemes also seem to have applicability as water quality criteria.

(3) The use of benchmark data in early stages of hazard assessment strategies was demonstrated in the conference, although concern was expressed with respect to the definition of the term "benchmark," as well as the intent of uses of benchmark information. Benchmarks are reference chemicals for which the toxic effects and environmental exposure (behavior) have been studied extensively. The use of this information through general comparisons with properties of untested chemicals in initial tiers can be used to guide the need for further testing. Benchmark data can also be used as decision criteria in hazard assessment, especially for very early identification of hazardous new chemicals.

(4) The workshop participants explored several possible options for employing decision criteria in hazard assessment schemes. These options include the following general considerations and conclusions:

(a) *Flexible versus rigid criteria*—Although early tier testing decisions are usually more rigid than later tier decisions, the concept of flexible criteria was preferred. This suggests that the margin between exposure and effect concentration may be acceptable for one compound and not for another. The flexibility of a specific decision criterion can be based on scientific judgment.

(b) *Volume and mammalian data*—Decision criteria should be more restrictive for compounds with high environmental release rates or adverse mammalian effects data. Neither parameter was quantified but should be as soon as possible.

(c) *Single versus multiple parameters*—It is desirable for each decision to be based on as many considerations as possible rather than the result of a single test.

(5) A fundamental assumption of the hazard evaluation approach is that laboratory fate and effect data can be extrapolated to field situations. However, inspection and extrapolation of the basic effect curve derived from a short-term LC50 test can give first indications of what is going to happen at low concentrations in long-term exposures. Validation of this assumption is difficult. In the few cases where exposure concentrations can be accurately determined, laboratory derived MATC values do predict effects to within a factor of two.

(6) Assessment of potential distribution in the environment of many chemicals at an early stage in their development is a goal for efficient evaluation of environmental hazards. Useful estimation of certain key physical and chemical property values has been made possible by use of equations derived from the available experimental data bases. Octanol/water partition coef-

ficients can be calculated on the basis of chemical structure alone. Such octanol/water partition coefficients can then be used through the derived equations to calculate a reasonable approximation of data on bioconcentration factors in fish, water solubility, and soil absorption coefficients, all useful hazard assessment tools. Mass balance distribution of a chemical between air, water, and soil can be calculated from its molecular weight, vapor pressure, and water solubility. These derived data, as well as those for bioconcentration, represent "worst case situations" since they do not include the effects of metabolic and degradative forces. These forces can be factored in later.

Based on a review of the literature concerning LC50's, MATC's, and Application Factors for fish and *Daphnia*, and the use of this data base, equations have been derived to predict these values for closely related species, mainly on the same compound. It was recognized that there is no general substitute for an experimentally derived MATC. Greater use could also be made of dose-effect time curves from short-term tests to predict effects of low concentrations in long-term exposures.

(7) Bioconcentration factors and uptake/depuration rates can provide useful information for assessing risk to consumers of aquatic biota of toxic substances. When used with mammalian, avian, and wildlife toxicity data, the relationship between toxicity and body residues of exposed organisms may provide additional information for evaluating the importance of field exposures. Bioconcentration factors should be considered with each new compound but actual bioconcentration studies need be performed only if persistence and octanol/water partition coefficients or solubility indicate bioconcentration potential.

(8) Aquatic hazard assessment procedures which are based on a well-defined and scientifically acceptable tier-testing systems are effective methods of determining the potential risk of toxic substances in the environment. However, it should also be stressed that present aquatic test procedures are designed to assess hazard to a limited class of aquatic organisms—fish and invertebrates. There is a need to develop a more comprehensive hazard assessment system which would include risk assessment procedures for aquatic mammals, birds, and plants.

In general, aquatic environmental fate and toxicological data must be viewed in the layered context of other environmental and health effects. For example, knowledge and understanding of terrestrial and human health data may often become an important factor to consider when analyzing the impact of bioconcentration of a toxic substance in the aquatic food chain. Similarly, preliminary knowledge of mammalian and avian biological effects data may provide guidelines for more detailed examination of effects on aquatic mammalian and predatory avian species.

(9) It was felt that some potential weaknesses in the current hazard assessment approach should be addressed by research of a general and fundamental nature of bioassay methodology. Thus the group concluded that there is a need to develop test procedures which reflect a chemical's impact on communities and on ecosystem functions such as detritus processing, photosynthesis, energy flow, chemical transforms, etc. In addition, it was felt that more emphasis should be placed on algal and plant tests and particularly under more realistic environmental conditions than heretofore. An important purpose of such research would be to examine the present approach which presumes that fish and other specific organism bioassay results provide an adequate basis for protection of aquatic ecosystems.

It was also felt that maximum possible use should be made of toxicity test relationships that exist from one species to another and from one kind of bioassay to another.

Finally, there was concern for the need to continue standardization of bioassay methods particularly with respect to environmental interactions, presence of suspended solids, diet, test medium impurities, etc.

Reference

CAIRNS, JR., JOHN, K. L. DICKSON, AND A. W. MAKI, EDS. 1978. Estimating the hazard of chemical substances to aquatic life. ASTM STP 657. Am. Soc. Test. Mater., Philadelphia. 278 pp.

WATER QUALITY CRITERIA

A Fisheye View of Water Quality Criteria

Kenneth J. Macek and Sam R. Petrocelli

*EG&G, Bionomics
Wareham, Massachusetts 02571*

Abstract

The process currently utilized to derive water quality criteria was characterized as to its objectives, the mechanism used to accomplish these objectives, the validity of the conclusions (i.e., criteria) derived, the level of risk judged to be unacceptable, the degree of (and need for) conservatism in the process, and the ecological relevance of the criteria derived.

In the opinion of the observers, the objectives are realistic and mechanisms are generally appropriate. It is suggested that the risk levels, the degrees of conservatism utilized, and the ecological relevance of the criteria derived with respect to certain of the objectives identified may be inappropriate.

The charge given to us by the organizing committee of this workshop in preparing our "discussion initiation paper" was threefold in nature. Two aspects of our task we will defer to the other eminent speakers in the workshop, while the third we will evaluate, at least philosophically, in some detail in an attempt to "initiate discussion."

The first task was to discuss "the necessary physical, chemical and biological data which, in our opinion, comprise the information necessary for the establishment of scientifically valid and ecologically relevant water quality criteria." We have chosen not to be redundant to the other speakers of this workshop who, during earlier presentations, have adequately addressed the "shopping list." With respect to the first charge, however, we would like to point out that although the terms "scientifically valid" and "ecologically relevant" may strike the reader as being synonymous, they, in fact, can be mutually exclusive. By way of analogy, we would suggest that the construction of a valid argument in science is similar to the construction of a valid syllogism in the study of logic. Given two fundamental (but erroneous) premises, one can easily construct a valid syllogism which is totally irrelevant to the real world. Surely we have all seen examples of precise application of the scientific method to a problem which yields a completely irrelevant conclusion relative to a solution because one of the fundamental assumptions (premises) was inappropriate. We believe we must constantly differentiate between "scientifically valid" and "ecologically relevant" for in deference to the former, it is clearly the latter which we must strive towards.

The second task with which we were charged was to "discuss the rationale for selecting the types of data" included in the above-mentioned shopping list. Again, we will defer to earlier speakers, and to the consensus reached at last year's workshop at Pellston, which have dealt with this subject in some detail.

The third task given us in our charge from the organizing committee, perhaps almost as an afterthought, was to discuss "our interpretation of the relative impact of chemical and biological data on the establishment of water quality criteria." It is this last subject to which we would like to direct your attention and hopefully discussions. We are going to assume that by now you are all familiar, at least in a qualitative sense, with the types of data with which we are here concerned. We are also going to assume that, to a lesser extent, you are familiar with the general processes by which such data are used to derive water quality criteria. What we would like to do is to anthropomorphize (in a piscitorial sense) for awhile and expose you to a fish's view of how we are doing in our attempts to establish ecologically relevant water quality criteria.

As fish we have observed, occasionally through rather murky water, that you humans seem to be experiencing a great deal of difficulty and controversy about what conclusion (i.e., criteria) should be derived from a particular toxicity data base. We have further observed that this confusion and controversy appears to be related to the fact that you cannot objectively and confidently measure either the experimental error (between species) or the sampling error (within species) in the toxicity data. This situation leaves you statistically infertile and dictates that the process be either highly subjective, artifically precise, or both. In both cases, it appears that you are left trying to juggle a data base of unknown statistical precision, with a therefore unknown risk associated with making an erroneous conclusion, with a therefore necessary degree of conservatism to balance the risk. Unfortunately, most of you humans are no better at juggling than we fish are at the backstroke, although you appear to do a much better job of it in some areas than others.

In order to explain our assessment of the situation, we would like to dissect and discuss the various aspects of the criteria-derivation process (Table 1). Let us for purposes of discussion consider those aspects of the process which are anthropocentric, those which are piscicentric, and those which might be described as ecocentric. We assume there is no need to define the first two terms. For purposes of this discussion, ecocentric shall refer to those concerns about all of the other biota (excluding we fish) in the aquatic environment, presumably as they relate to the functioning of the aquatic ecosystem as a whole. We would like to look at each of these aspects from the standpoint of the objectives of the criteria, the adequacy of the mechanisms used to derive the criteria, the relative risk of being wrong in your conclusion (criteria), and the degree of conservatism you use to compensate for the relative risk you perceive to be associated with error.

Discussion

The anthropocentric objectives of water quality criteria appear to be the most important to you and are primarily concerned with protecting all individuals of a single species (humans) from possible adverse effects associated with the occurrence of chemicals in water. The two most obvious threats are toxicity due (1) to drinking

TABLE 1.—*Rationale and purpose of the criteria-derivation process.*

	Anthropocentric	Piscicentric	Ecocentric
Objectives	Drinking water quality. Food resource quality. Recreational value.	Protection of fish populations.	Protection of all other populations of aquatic organisms.
Mechanisms	Mammalian toxicology. Bioconcentration testing. Mutagenicity testing. Teratogenicity testing. Carcinogenicity testing. Extrapolation from lower biological systems to man.	Acute toxicity tests. Short-term tests to estimate chronic toxicity. Chronic toxicity tests. No extrapolation. Extrapolation from one biological form to other closely related forms.	Acute toxicity tests. Some actual chronic toxicity tests. Emphasis on most sensitive forms. High degree of extrapolation from some forms to other not necessarily related forms.
Risk of Error	Loss of individual(s) within a single species.	Loss of individual(s) within a group of closely related species.	Loss of individual(s) within a group of diverse forms.
Conservatism	Extremely high apparently due to level of acceptable risk and degree of extrapolation required.	Relatively high for reasons not readily evident.	Extremely high apparently due to extrapolation gap.

the water or (2) eating the food resources contained therein. The procedure used to protect humans include the use of classical mammalian toxicology mechanisms and, more recently, mutagenicity, teratogenicity, and carcinogenicity studies to extrapolate these data from lower forms to humans, and thus to estimate presumed acceptable daily intakes for humans. These then are related to concentrations of the chemical in water which are acceptable from a direct consumption standpoint (drinking water) and to concentrations in water which will limit residues in food resources to acceptable levels. This latter relationship is derived by measuring the bioconcentration process in aquatic organisms. All of these mechanical approaches have been traditionally formalized and rigorously utilized, seem to have worked fairly well, and are generally accepted in the scientific community as valid and relevant. Because the unacceptable risk of error is measured in terms of the loss of one or a few individual organisms in the population (truly an anthropocentric perception of risk) and because of the inability to measure validity or precision of the extrapolation process over a relatively large biological gap, there usually is applied what appears to us to be a relatively high degree of conservatism to the process.

However, there appears to be little controversy among you humans about whether such conservatism is required, and in that sense the degree of conservatism utilized appears to be appropriate for the level of risk considered unacceptable. We consider it pure folly to comment on the ecological relevance of protecting all individuals in a population as we are sure your view of our ecosystem, and the importance of your niche, differs greatly from ours. We should point out, as it will come up again in this discussion, that little consideration appears to be given to variations in sensitivity among the population which may be due to race, nationality, sex, age, life stage, etc. The concept of a man is a man is a man (at least in a toxicological sense) appears to be universally accepted, and there does not appear to be any evidence that catastrophic events have occurred as a result of the universal application of this concept.

Recognizing that we might be accused of suffering from myopia, we suggest that the piscicentric objectives of water quality criteria are the most important. We observe that you appear to be concerned (for whatever motives) with protecting all individuals of a variety of closely related species from direct toxic effects associated with the occurrence of chemicals in water. The processes used to protect us are comparable to classical mammalian toxicology approaches. That is, you attempt to empirically develop an understanding of the acute toxicity and chronic toxicity of the chemicals to fish, extrapolate from empirical observations on some of us to the rest of us, and attempt on the basis of empirical information to estimate non-toxic concentrations in water. The acute toxicity tests, critical life-stage tests, and full chronic tests appear to be reasonably good mechanisms for measuring sensitivity of fishes to chemicals, and the conclusions about toxicity derived therefrom appear to be valid. The reasons for this include the facts that you appear to have a great deal more knowledge and experience in collecting, handling, maintaining, and working with us than any other aquatic group. Furthermore, contrary to the situation inherent in mammalian toxicological testing, in which results from tests with sub-human mammals are extrapolated to man, fish are tested directly and the effects of chemicals are frequently assessed on those species, or closely related ones, which are most likely to be exposed to these chemicals. Clearly, extrapolation from one fish species to another involves a much smaller potential risk of error than does the extrapolation from a bacterium or rodent to humans.

For reasons unclear to us, you seem here again to be concerned with measuring the unacceptable risk of error in terms of loss of one or a few individuals in the population (since concentrations producing any significant deviation from control performance during a test are considered unacceptable) rather than emphasizing population effects which would be ecologically relevant. We question (because we can apparently be more objective than you) the ecological relevance of criteria designed to protect all individuals of few closely related and generally prolific populations. Finally, there appears to be an inappropriate amount of conservatism built into the process when one considers that the "extrapolation gap" is relatively small. We should point out that despite the emergence of a fairly impressive data base to support the recently suggested concept that a fish is a fish is a fish, you seem fairly reluctant to accept and apply the concept, as evidenced by your emphasis on variability in sensitivity due to race, nationality, age, sex, life stage. For our part, except for the occasional individual with well-endowed pectoral fins and/or a particularly attractive heterocercal tail (pardon the anthropomorphism), we would be inclined to agree that a fish is a fish is a fish, particularly in relation to thresholds of chronic toxicity.

The ecocentric objectives of water quality criteria appear to be primarily concerned with protecting all individuals of all other species from direct toxic effects associated with the occurrence of chemicals in water. The process used to protect our fellow citizens of the aquatic environment are again related to understanding the acute and chronic toxicity of the chemical to certain of the inhabitants, estimating threshold concentrations of chronic toxicity, extrapolating over a wide variety of diverse biological forms, and applying a considerable degree of conservatism to protect all individuals in all populations. The mechanisms are the same as those used to accomplish the piscicentric objective; however, we question whether, when used here, they are as accurate and precise in measuring sensitivity of organisms to chemicals. It appears to us that there could be a high degree of sampling error associated with the use of these mechanisms here, and certainly we must be concerned about the lack of information regarding experimental error. The variances which are observed in data resulting from toxicological studies with these organisms may in reality reflect more of our ignorance of appropriate maintenance and testing techniques than any extremes in sensitivities to chemical exposure. Furthermore, in current practice, results of tests with only a few species, some of which are relatively insignificant ecologically in the environment, are extrapolated across large phylogenetic gulfs to groups of organisms which are quite diverse developmentally, apparently in the attempt to protect nearly all, if not all, species at all times. Thus, we could reasonably question the validity of the data and the conclusions derived therefrom.

Furthermore, we are indeed perplexed by the establishment of risk levels such that all individuals in all populations should be protected all of the time. Regarding the ecological relevance of such criteria, if, as we presume, you are primarily interested in the functioning of the ecosystem as whole, then surely you are aware that there exists a tremendous reproductive potential in most species of aquatic organisms, and that it is folly to equate the risk of losing a few individuals of some lower aquatic species, with the risk of, for example, losing some individuals from a fish population (or even the human population, for that matter). Furthermore, you must be aware that there are a number of lower aquatic forms which are capable of filling each functional and structural niche in an ecosystem, and that the loss of even an entire population of a species, in all likelihood, would not result in dramatic functional or structural changes in the ecosystem. For our part, as much as we appreciate the aroma of a serving of fresh zooplankton, or the delicate taste of midge larvae, we too believe variety in a diet is a desirable, if not an essential, characteristic. Furthermore, we are not particularly concerned about who lives next door or schools around with our offspring. Thus, we do not understand why you are obsessed with maintaining the status quo or, conversely, precluding any changes in our neighborhood.

Lastly, we do not understand your obsession with piling conservatism upon conservatism at every step of the process to the point of absurd redundancy. We recognize that all of this conservatism is motivated by your desire to compensate for all of the "apparent differences in sensitivity indicated by the results of your toxicity studies with lower aquatic forms." We would suggest that in your typical anthropocentric manner you are interpreting all of the "spread" or "noise" in your toxicity data as a measure of the range of sensitivity which exists among lower aquatic forms. We wonder if you have asked yourselves how much of the "noise" or "spread" in your data is due to sampling error (i.e., your ability to accurately measure the sensitivity of any species) or to experimental error (i.e., your ability to recognize the representativeness of species tested). It appears to us that you may be using all of this obvious conservatism to compensate for perceived differences in sensitivity caused by inadequate tools.

Summary and Conclusions

We have attempted to summarize in matrix form our objective analysis of the key characteristics of your criteria-setting process (Table 2). In addition, we are providing in matrix form our

TABLE 2.—*Objective analysis of key characteristics of the criteria-setting process.*

	Anthro-pocentric	Pisci-centric	Eco-centric
Objectives	Reasonable	Admirable	Questionable
Adequacy of mechanisms	Good	Excellent	Unknown
Appropriateness of risk level established as unacceptable	Moot	Unrealistic	Absurd
Degree of conservatism applied	Very high	Too high	Extremely high
Ecological relevance	Moot	Unreasonable	Unrealistic

opinion of the appropriateness of the objectives involved, the adequacy of the mechanisms, the appropriateness of the risk judged to be unacceptable, the degree of conservatism involved, and the ecological relevance of the results derived from the process as we currently perceive it.

In summary, we are not optimistic that our comments on any of the anthropocentric aspects of the criteria-derivation process will rise above the pleuston, be picked up by your auditory mechanism, and stimulate meaningful discussion. We fully expect to have these comments dismissed with a reaction similar to "What do they know, anyway?" In addition, despite the fact we consider ourselves somewhat of an expert on the subject, we will not even be too disappointed if our comments on the piscicentric objectives are dismissed with some thought such as "What would you expect them to say!".

However, we sincerely hope that our characterization of the ecocentric objectives will cause you to think again about all aspects of the criteria-derivation process related to lower forms, and to the ecosystem as a whole. The critical concern relative to stable populations of other aquatic species should be the maintenance of functional species diversity such that the presence of specified concentrations of chemicals in water will not result in the selective elimination of a particular functional group (e.g., trophic level) from the aquatic community. This is, shifts in species diversity in aquatic communities is not necessarily detrimental if the species represent all of the appropriate functional types.

Furthermore, to be honest, we really are getting tired of all of the construction (of treatment systems) going on in the neighborhood. Much of the construction is unnecessary, and some create more problems of a greater ecological impact on us (not to mention economic impact on you) than those problems which they are intended to correct.

A Proposed Method for Deriving Effluent Limits from Water Quality Criteria

JERRY L. HAMELINK

*Toxicology Division
Lilly Research Laboratory
Box 708, Greenfield, Indiana 46140*

Abstract

A method is described for determining when and for what duration the maximum recommended concentration of potentially toxic chemicals can be allowed to exist in a river to ensure protection of aquatic life. The concentration of a chemical arising from a point source in a river is dependent upon three factors: (1) the volume and hydrodynamic characteristics of the water flowing in the river; (2) the quantity of chemical being discharged; and (3) the kinetics of environmental fate reactions for the chemical. A probabilistic approach is outlined for determining the likelihood of these three factors exceeding some predetermined limits. By applying this approach, site-specific water quality standards which permit consideration of probable dilution volumes, hydrodynamics, environmental fate reactions, process variabilities, and other factors can be developed. An example of how the method works is given for the Wabash River at Lafayette, Indiana receiving a point source discharge from a hypothetical plant manufacturing a hypothetical chemical.

The Environmental Protection Agency (EPA) is proceeding with the establishment of water quality criteria for a number of chemicals. For each of the pollutants so identified "the criteria are to state maximum recommended concentrations consistent with the protection of aquatic life and human health" (Water Quality Criteria 1978). The purpose of this discussion is to consider when and for what duration the maximum recommended concentrations of potentially toxic chemicals might be allowed to exist and when variations in concentration might be appropriate in natural waters.

Rationale

Concentration is the cornerstone of all aquatic toxicity measurements. From laboratory studies we can determine that concentration of a chemical over a selected period of time which will or will not result in some measureable toxicity or tissue residue. Now we need to consider what concentration *for what period of time* will be acceptable or unacceptable in a natural river.

The concentration of a chemical arising from a point source will be dependent upon three factors: (1) the volume and hydrodynamic characteristics of the water flowing in the river; (2) the quantity of chemical being discharged; and (3) the kinetics of environmental fate reactions for the chemical. None of these factors are constant, but they should operate within definable limits of space, mass, and time (i.e., definable as a probability statement for the likelihood of exceeding some limit). Hence, this discussion will be devoted to some of the factors that might be used to define the limits and duration of exposure for a toxic chemical, as opposed to an oxygen demanding substance, given a hypothetical set of water quality criteria values.

Methods

River Flow

We normally think of the discharge of a river as a series of high and low stages which encompass an occasional spring flood and a periodic summer drought. Most of the time, the river's flow is somewhere in between the two extremes. The problem is, we can never know when the level of water in the river will be high, or low, or somewhere in between. To circumvent this difficulty, hydrologists have developed probability tables for flows in individual rivers based on historical records (Rohne 1972; Horner 1976).

Hydrological probability data are site-specific. For purposes of this discussion I have selected the Wabash River at Lafayette, Indiana (Table 1). Similar frequency-of-occurrence data are available for most rivers throughout the US. When the *low flow* data for this station for all years of record combined are analyzed for days of duration, a simple relationship between the median discharge and duration of occurrence is evident (Figure 1). For purpose of extrapolating laboratory data, the minimum probable flow volume available for dilution was chosen to be the lower ca. 90% tolerance limit as determined according to the method of Somerville (1958). Hence, it can be stated that 90% of the time we can be confident that 95% of the future discharge volumes will exceed the lower discharge volume shown for a corresponding period of time. The median low flow discharge for one day is 22.48 cms ranging upward to 30.61 cms for 30 consecutive days for all years combined. The corresponding 90% lower boundry tolerance limits are 13.36 cms for one day and 15.63 cms for 30 consecutive days. The 7-day, once-in-10-years summer low flow, which is often used for planning programs concerned with the continuous discharge of oxygen demanding substances, is 16.71 cms (Rohne 1972).

Quantity from Source

The quantity of an individual chemical that will be discharged from a manufacturing facility is not constant. However, like the river, the frequency and duration of discharge can be defined for specific chemicals from specific sites. In essence, 24-hour composite samples of effluent are routinely collected, and assayed for the chemicals of interest. By concurrently measuring the volume discharged, the daily load can be determined and the variability in the release rate assessed. Experience has shown that data from these sources usually fit a *log-normal* distribution more closely than a normal distribution. Thus, the variability expressed as standard deviations from the mean of the logarithms represents a *ratio* of *two discharge loads,* where the mean constitutes a *median* value (i.e., 50% probability of occurrence) for the sampled population and the daily limits are selected probabilities of occurrence (e.g., the maximum daily load approximates the ratio of the 99% to the 50% probability of occurrence, or two standard deviations from the measured median). Consequently, variability in the discharge rate under the National Pollutant Discharge Elimination System (NPDES) program is normally limited to maximum daily load and an average daily load for the maximum month. The average daily load for a year can also be determined, but this is not practical for enforcement programs.

Probability distributions of this nature have been used to establish effluent, or so called end-of-pipe, guidelines. For example, the Best Practical Treatment (BPT) effluent guidelines for industry point source categories recognize the inherent fluctuations in wastewater treatment plant operations and production schedules by incorporating variability factors into each set of performance standards (PMPSC 1976; PCMPSC 1976). For discussion purposes, a conservative variability ratio of 4:2:1 for the quantity discharged on the maximum day:average day for the maximum month:average annual day, respectively, will be used.

TABLE 1.—*Duration of daily flows in the Wabash River at Lafayette, Indiana for indicated periods of time (from Rohne 1972).*

Months	Period	98%	95%	90%	80%	70%	50%	20%	10%
3	Aug.-Oct.	14.72	17.56	20.11	25.20	30.01	40.21	74.48	113.27
6	May-Oct.	16.42	19.54	23.22	31.4	39.65	62.58	144.42	237.88
3	June-Aug.	16.71	21.52	26.62	35.11	44.18	68.24	137.34	218.05
12	Oct.-Sept.	18.41	22.65	28.32	37.95	49.56	82.41	226.54	399.29

Discharge, in cms, which was exceeded for indicated percent of time during 1925–66 water years

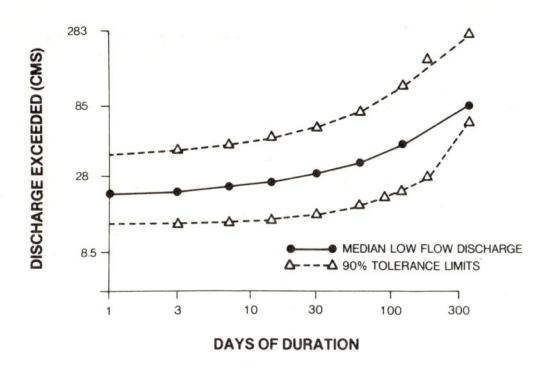

FIGURE 1.—*Median and 90% tolerance limit distribution of the lowest mean discharge (cfs) volume of the Wabash River at Lafayette, Indiana for consecutive days in a year ending March 31 for 40 years (1924–1973), combined.*

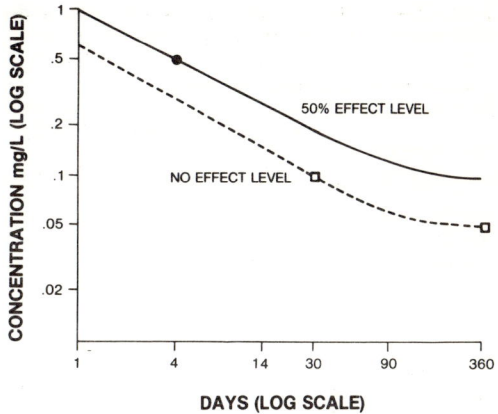

FIGURE 2.—*The time-dose relationship of a hypothetical chemical that resulted in a toxic effect on 50% of a population of organisms (EC50) or caused no observable effect (EC0).*

Environmental Fate and Hydrodynamics

The environmental fate reactions of a chemical and the hydrodynamic characteristics of a receiving body *cannot* be generalized. Therefore these factors must be considered for each individual chemical on a site-specific basis.

Toxicity Information

The occurrence of a toxic effect or no-effect on an aquatic species depends on both the concentration and duration of exposure. In order to provide a quantitative basis for our discussion, I have constructed a pair of toxicity curves (Sprague 1969) for a hypothetical chemical (Figure 2). (NOTE: Any relationship that exists between the slope of the curves, the range between effect or no-effect, and the concentration values used to construct the hypothetical case with the EPA's Water Quality Criteria 1978 is strictly intentional.) A similar toxicity profile could be drawn for any chemical for which acute, 30-day and chronic EC50, and no-effect data are available. It is recognized that certain life stages may be more sensitive than others to brief periods of exposure. Hence, the toxicity profile refers to the duration of exposure required to elicit an effect and not to the amount of time required to complete a particular test.

Results

Given the river flow, source quantity, and toxicity information, the "maximum recommended concentration consistent with the protection of aquatic life and human health" may be derived several different ways. For the sake of brevity, the variety of possible alternatives will be limited to three examples.

Strategy #I

The criteria might be applied by assuming that the chronic no-effect concentration (i.e., 0.05 mg/liter) would exist if the maximum daily load were diluted by the summer low flow (i.e., 7-day, 10-year low = 16.71 cms) of the river (Rohne 1972). The corresponding point source discharge limitations would then be established by the discharge variability distribution as ½ and ¼ of the daily maximum (Table 2). Under this strategy, the maximum daily discharge allowance would be 72.3 kg and the daily average for the maximum month would be 35.9 kg.

Strategy #II

The annual average daily load might be derived based on the final chronic no-effect con-

TABLE 2.—*Discharge loads (kg/day) allowable under Strategies I, II, and III for a hypothetical chemical, for possible enforcement sampling intervals of time, assuming a 4:2:1 variability ratio.*

Sampling Interval	Strategy I (kg/day)	Strategy II (kg/day)	Strategy III (kg/day)
Maximum day	72.27	1425.45	694.09
Average day, maximum month	35.90	712.72	136.36
Average annual day	18.18	356.36	90.91

centration value by assuming it was diluted by the annual median low flow (i.e., 50% Oct.-Sept. = 82.41 cms) of the river (Rohne 1972). The daily average for the maximum month and the daily maximum load limits would then be defined as 2 and 4 times and average annual daily load, or 712.7 kg and 1,425.4 kg, respectively (Table 2).

Strategy #III

Point source discharge limitations might be derived independently of production variability by "cross multiplying" river flows for selected durations of time, in accord with desired probabilities of occurrence (Figure 1), with the no-effect toxicity levels for the same time periods (Figure 2). The quantity, in kilograms, that would be permitted to be released into the river could be determined based on either the median discharge curve or the 90% lower boundry tolerance limit curve. In order to reduce the risk of error to a minimum, the 90% lower boundry tolerance limit curve was used to set the hypothetical effluent load limits (Figure 3) [e.g., 24-hour EC0 = 0.6 mg/liter × (13.36 cms · 2.45 × 10^6 kg H_2O/day/cms) = 694.1 kg/day]. (NOTE: The actual release rate could not follow this curve because the average day for the maximum limit would be exceeded by a factor of ca 2.) Thus, for this strategy the maximum daily allowance would be 694.1 kg and the daily average allowance for the maximum month could be reasonably set at 136.4 kg/day and the annual average daily load might be reduced to 90.9 kg/day. (NOTE: In theory, the annual average daily load could be allowed to range up to 295.4 kg/day without posing a significant risk to the population) (Table 2).

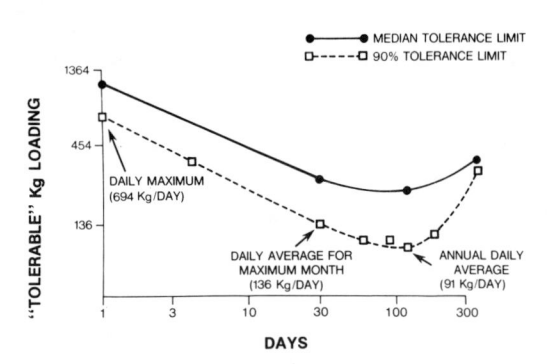

FIGURE 3.—*Amount of hypothetical chemical that could theoretically be tolerated by a population of aquatic organisms in the Wabash River below Lafayette, assuming worst-case conditions.*

Discussion

Strategy #I is the most conservative and least realistic method presented. Even if the maximum daily load and the day of minimum flow occurred on the same day, it ignores the duration of exposure required to elicit an effect, as presented in Figure 2. It is admittedly difficult to imagine that the concentration of any toxicant shown to be "safe" for an entire life cycle could be deleterious to a natural community if it existed for only one day every year. Thus, this strategy is rejected for being unrealistic.

Strategy #II would be appealing to those who discharge effluents because it incorporates the aspect of variability inherent in release rates from a point source, yet assures that the aquatic community would be protected on a chronic basis. The difficulty with this strategy is it fails to take into account short-term toxicity. That is, by arbitrarily increasing the maximum monthly and daily allowances, a significant risk of encountering acute effects may arise. In reality, this "risk factor" would be dependent upon both the slope of the toxicity curve and the actual discharge ratios. Hence, the "risk factor" could be determined and reduced to an acceptable level. However, it would then become a difficult strategy to administer because the effluent discharge ratios would have to be established based on experience. Thus, this strategy would only be practical for toxicants for which a discharge history was available.

Strategy #III should be the most appealing to toxicologists because it is derived from the premise that a finite duration is required to elicit an effect or, conversely, some concentrations can be tolerated without harm for a short period of time that would be deleterious over a longer period of time. Secondly, it assures that the duration of exposure to a concentration likely to produce an effect has a low probability of occurrence.

Consideration of this strategy reveals one other rather startling observation that, for lack of a better definition, I have termed "the paradox of the parabola." Because the low-flow period is usually less than 30 days duration, the probable dilution available to protect against chronic effects nearly compensates for a 2× lower concentration between the 30-day EC0 and the full-life cycle EC0. Coupled with the fact that most river fishes spawn during the spring high-flow period, the life stages that are usually most sensitive to toxicants naturally receive further protection. Consequently, 30-day (e.g., egg-larval) tests (McKim 1977) would appear to provide sufficient information for the establishment of water quality criteria to protect aquatic life and also provide a convenient means for implementing the criteria into site-specific standards.

Conclusions

(1) Water quality criteria to protect aquatic life have to be derived for each individual chemical based on toxicity studies that provide a measure of both the EC50 and EC0. Numerical "application factors" should only be used as a general guide when definitive information is lacking.

(2) Water quality standards developed from the criteria must be site-specific. This will permit consideration of probable dilution volumes, hydrodynamics, environmental fate reactions, process variabilities and other factors that cannot be derived in the laboratory.

(3) Toxicity tests of 30 days duration probably provide sufficient information to develop water quality standards for point sources discharging to rivers.

(4) Probable duration-of-exposure relationships for other potential sources and aquatic ecosystems should be investigated as a simple technique for reducing complex problems in hazard evaluation to an understandable and practical scale.

References

HORNER, R. G. 1976. Statistical summaries of Indiana stream flow data. Pages 158–160 in an open file report prepared by U.S. Dep. Inter., Geol. Surv., Water Resour. Div. in cooperation with Indiana Dep. Nat. Resour., Div. Water.

McKIM, J. M. 1977. Evaluation of tests with early life stages of fish for predicting long-term toxicity. J. Fish. Res. Board Can. 34:1148–1154.

PCMPSC (PESTICIDE CHEMICALS MANUFACTURING POINT SOURCE CATEGORY)—Part 455. 1976. Fed. Regist. 41(211):48088–48096.

PMPSC (PHARMACEUTICAL MANUFACTURING POINT SOURCE CATEGORY)—Part 439. 1976. Fed. Regist. 41 (223):50676–50686.

ROHNE, JR., P. B. 1972. Low-flow characteristics of Indiana streams. Page 115 in an open file report prepared by U.S. Dep. Inter., Geol. Surv., Water Resour. Div. in cooperation with Indiana Dep. Nat. Resour., Div. Water.

SOMERVILLE, P. N. 1958. Tables for obtaining nonparametric tolerance limits. Ann. Math. Stat. 29:599–601.

SPRAGUE, J. B. 1969. Measurement of pollutant toxicity to fish, I. Bioassay methods for acute toxicity. Water Res. 3:793–821.

WATER QUALITY CRITERIA. 1978. Fed. Regist. 43(97):21506–21518.

The Development of Water Quality Criteria—
An Environmentalist's Viewpoint

DANA J. DAVOLI AND EILEEN R. CHOFFNES

Citizens for a Better Environment
Chicago, Illinois 60605

Abstract

In order to develop water quality criteria (WQC), a data base must first be compiled for each pollutant. This data base must include all the relevant information available on the pollutant—including its physical and chemical properties, methods of transport and ultimate fate in the environment, and its effect on a wide range of organisms with which it might come in contact. The criteria formulated from such a data base must be conservative enough to account for the inherent unknowns in our understanding of toxicology, the difficulty of extrapolating data from laboratory studies to the real world, and the myriad toxins to which all organisms in our environment are exposed.

The Environmental Protection Agency's (EPA) development of water quality criteria for the 65 pollutants listed in the Consent Decree are inconsistent with these philosophies and, as such, will not result in WQC that will protect man and the ecosystem at large.

In order to develop water quality criteria (WQC), a data base must first be compiled for each pollutant. This data base must include all the relevant information available on the pollutant—including its physical and chemical properties, methods of transport and ultimate fate in the environment, and its effect on a wide range of organisms with which it might come in contact.

This data base must then be evaluated so as to decide the maximum level of the pollutant permissible that will still protect the public health and maintain a balanced ecosystem. Where appropriate, this maximum concentration might be zero.

In the following discussion we will focus on the type of information that must, when available, be included in any data base used to establish WQC. We will then state our position on how such criteria should be formulated. Finally, we will comment on EPA's development of WQC for the 65 pollutants designated in the Consent Decree, to which Citizens for a Better Environment (CBE) is a party.

Discussion

The Data Base

Table 1 lists the type of information that must be included, when available, in the data base provided in a criterion document. This information is critical to proper formulation of such a document.

Physical and Chemical Properties and Environmental Fate and Transport

To estimate the level of exposure of susceptible organisms to a water pollutant, it is important to study the movement of the pollutant from its manufacturing site to its ultimate reservoirs—both physical (water, sediment, and air) and biological (living things). We must know the potential sources of contamination, both point and non-point, and the amount of the pollutant discharged. Any impurities in the commercial-grade products must also be known,

TABLE 1.—*Information to be included in the data base used in formulating water quality criteria.*

Physical and chemical properties—including, but not limited to:

<u>Nomenclature</u>—common name and chemical name
<u>Chemical structure</u>—possible isomers and/or salt forms
<u>Description</u>—chemical group, color, odor, liquid or solid
<u>Specific chemical and physical properties</u>—molecular weight, melting point, boiling point, vapor pressure, solubility (in polar/non-polar solvents), partition coefficient (octanol/water), sorption (soils and sediments)
<u>Origin</u>—natural or synthetic, origin and uses, general method of synthesis or production, chemical composition and impurities of commerical grades
<u>Pertinent chemical reactions</u>—photo-oxidation, microbial oxidation, salt formation, formation of organometallic compounds or complexes, auto-oxidation, pH effects, effect on other chemicals in water

Environmental fate and transport—including, but not limited to:

<u>Potential sources of contamination</u>—point sources and non-point sources
<u>Concentrations present in soils, waters and tissues</u>—industrial and municipal effluents, soils, sediments, natural waters, air, drinking water sources, human food supply, tissues of aquatic and terrestrial organisms
<u>Persistence in soils, water, and air</u>
<u>Biodegradation and biotransformation</u>—photolysis, microbial degradation, formation or degradation of organic-metallic compounds or complexes, etc.
<u>Bioconcentration, bioaccumulation, biomagnification</u>—concentration in food or water, time period, species and concentration factor

Toxicological and metabolic data

These data should include information on microorganisms (aquatic/non-aquatic), plants (aquatic, including phytoplankton and algae, and non-aquatic), invertebrates (aquatic/non-aquatic), vertebrates (aquatic and terrestrial including fish, birds, non-human mammals, and humans).

<u>Acute/subacute toxicity</u>
<u>Chronic toxicity</u>—mutagenicity, teratogenicity, carcinogenicity, neurotoxicity, etc.
<u>Other toxicological or metabolic effects</u>—biochemical changes, behavioral effects, ecosystems effects, etc.
<u>Pharmacokinetics</u>—absorption, distribution, metabolism, excretion
<u>Epidemiological data</u>

that will eventually arise in sediments, water, air, and in the tissues of aquatic and terrrestrial organisms depend, of course, on the persistence of the chemical in various media, as well as the alterations to which it is subject in the environment through processes such as photolysis, microbial degradation, auto-oxidation, and so forth. Concentrations found in organisms will also depend on the ability of the chemical to bioconcentrate, bioaccumulate, and biomagnify.

Very little attention has been paid to the fate of chemicals in the environment. Thus, for new chemicals and for most existing chemicals, little quantitative (or even qualitative) information is available on levels that arise in physical and biological systems. As a result, heavy reliance must be placed on available physical and chemical data on the pollutant as a means of predicting environmental fate and thus environmental exposure.

Information about the structure of the pollutant and its vapor pressure, solubility in polar and non-polar solvents (including octanol/water partition coefficients), and adsorption/desorption to particulates can help us to predict what levels of the pollutant will arise in water, sediment, and air once it is discharged. Known chemical reactions, both physical and biological, can indicate the stability of the chemical in the environment, as well as its ability to form salts and organometallic compounds or complexes. In addition, the structure of the compound, its resemblance to actively accumulated compounds, and the relation between fat and water solubility can be used to make a reasonable prediction of potential bioconcentration and bioaccumulation.

Toxicological and Pharmocokinetic Data

As Table 1 indicates, the data evaluated in developing WQC must include the effects of a pollutant on *all* organisms, both aquatic and non-aquatic, including microorganisms, plants, and all animals, invertebrate and vertebrate. This data base must include all available information on acute, subacute, and chronic toxicity and on pharmacokinetics. Unfortunately, for at least two reasons, such a comprehensive data base is rarely compiled—even when the necessary information is available.

One reason is that most decisions concerning environmental quality have been and are made

since the impurities may be more harmful than the chemical itself. For example, certain pesticides contain impurities such as dioxin and nitrosamines.

Once a pollutant is discharged, its movement into physical and biological systems can be estimated from its physiochemical properties or can be measured directly. Levels of the pollutant from an anthropocentric viewpoint. This is exemplified by the considerations most often addressed in the development of WQC: the protection of human health when water is used for domestic supply; the protection of human health when water is used for harvesting fish or shellfish for human consumption; and the maintenance of aquatic organisms that are economically or asthetically important to man.

Such an anthropocentric approach is totally unacceptable, however. Any discussion of the quality of life must address itself not only to the quality of human life but also to the quality of life of all other organisms in the environment, including microorganisms, vegetation, and other animals. Therefore, any toxicological data base must include, when available, information on a pollutant's effect on *all* organisms in the environment, not only on man or the organisms upon which he depends.

A second factor is that the toxicity information on which most health and environmental standards are based is very limited, even when extensive data are available. While most available acute toxicity information is always included in an analysis, chronic toxicity data are usually limited to a few effects on a limited number of organisms.

For example, in the regulation of human health hazards, most agencies, including the US Environmental Protection Agency, have focused their attention on carcinogens. But, while it is certainly important to achieve stringent control of carcinogens, other toxic chemicals such as mutagens, teratogens, and neurotoxins deserve equally as much attention. A birth defect caused by exposure to a mutagen or teratogen, or nervous system damage resulting from exposure to mercury or other neurotoxins is just as irreversible as cancer and just as distressing to the person afflicted. In addition, while it is widely accepted that there is no "safe" level of exposure to a carcinogen, this is quite likely also true for mutagens and possibly for many teratogens, especially those acting at the level of the genetic material (for example, alkylating agents). And the most recent studies of the effects of lead on young children challenge the assumption that there may be "safe" exposure levels to neurotoxins. These studies have demonstrated that the supposedly "low" levels of lead in our environment are sufficient to result in learning disabilities and other subtle neurological and behavioral effects (Needleman and Piomelli 1978).

If WQC are to protect human health and the environment, they must be based not only on acute effects data but also on everything that is known about carcinogenicity, mutagenicity, teratogenicity, neurotoxicity, and systemic, behavioral, and biochemical effects.

Criterion Formulation

Once a data base has been compiled, a consistent methodology must be applied in formulating water quality criteria. This methodology will be different for pollutants whose toxicological effects are believed to be threshold in nature, as opposed to those pollutants for which there is no known threshold of activity.

For the known non-threshold pollutants (carcinogens and mutagens), for which no safe level of exposure can be determined, WQC should be established at a level that will not increase the incidence of cancer in the United States. Estimates of risk for carcinogenic substances should be made using the linear non-threshold model. Estimates of carinogenic risk from mutagenic substances can be developed by extrapolating from mutagenicity tests such as the Ames test where mutagenic and carcinogenic potency have been correlated.

For pollutants whose toxicological effects are believed to be threshold in nature, WQC are normally derived from the results of toxicity tests (acute, chronic, MATC) by means of application factors and safety factors. It is assumed that such factors will result in toxicant concentrations that will protect human health and the environment. But while we support such an approach, too often WQC developed by such a method are not protective because the following problems are not taken into account:

(1) As we increase our understanding of the detrimental effects of chemicals on humans and other organisms, certain effects such as acute and chronic intoxication are more clearly perceived. Yet other effects are only now becoming apparent—for example, subtle effects such as the induction of cancer, modifications of neurological behavior and biochemical proc-

esses, and effects on reproduction. It may be years before we know the ultimate results of exposure of genetic material to teratogens and mutagens.

(2) Most of our toxicological data come from experiments on laboratory animals. Since such experiments are conducted under conditions that are optimal for the test organisms, they tend to minimize values for the toxicity of a tested substance. In the "real" world, organisms are often not under optimum conditions. In particular, they are not exposed to one toxicant at a time, but rather to myriad chemicals, many of which will exert additive and synergistic effects with one another (National Academy of Sciences 1973; Marking 1977).

(3) There are enormous problems in extrapolating from experimental data to actual environmental situations because of the differences between sensitivity in test species and sensitivity in organisms to be protected. For humans this problem is exemplified by thalidomide, to which humans are 60 times more sensitive than are mice, 100 times more sensitive than are rats and 700 times more sensitive than hamsters (Epstein 1974). Even within the human population, we must protect sensitive individuals such as children, unborn fetuses, persons with certain metabolic disorders, and the elderly.

For determining WQC that will protect organisms in the ecosystem other than man, we usually must rely on laboratory data compiled for species—"representative" from an anthropocentric viewpoint—that have been selected in part because they are well adapted to the laboratory environment. Such experimental data rarely yield information on organisms in the environment that may be more sensitive or that may already be endangered; nor do they yield information on organisms that are particularly vulnerable because of their position in the food chain, such as carnivorous birds whose entire diet may consist of aquatic life. And finally, such experiments do not yield information on the effect of a toxicant on interspecies or community interactions. Yet recent experiments with PCB's and other toxic substances have shown that community structure and interspecific competition in multispecies systems can be affected by concentrations of toxicants much lower than those required to show effects on a single species (Fisher 1974; Hansen 1976; Burks and Wilhm 1977).

Therefore, in deriving WQC, it is important to apply conservation safety factors that will take into account the inherent unknowns in our understanding of toxicology, the difficulty of extrapolating data from laboratory studies to the real world, and the myriad toxins to which all organisms in our environment are exposed.

One additional comment: It is often argued that conservative WQC require the installation of expensive control technology. It is becoming increasingly clear, however, that the cost of *not* controlling pollution will be much higher than will be the cost of controlling pollution through stringent regulations.

The economic impact of cancer, for example, is astronomical. Indirect costs of cancer to the victim, family, and community include loss of earnings during illness, loss of future earnings, and elimination of workers with skills and experience. Direct and indirect costs, such as treatment, hospital care and death costs, have been estimated at $19–28 billion in 1974 (Rice and Hodgson 1978).

Costs resulting from the effects of a teratogenic agent are likely to be greater than are those for cancer. The cost of caring for an individual with a birth defect will vary according to the severity of the defect. But, in the worst case, in which an individual needs constant care outside the home, the cost can be $36 thousand a year for custodial care in a state institution (T. Monahan, Chicago Associaton for Retarded Children, personal communication), or several hundred thousand dollars per year for private care. Over the life of the individual, the cost will be enormous. Even for those persons with less serious defects, money must still be spent for prosthetic devices and other forms of care such as special education. Similar estimates of cost can be made for persons suffering from neurological and systemic problems, many of which may be a direct result of environmental pollution (US Department of Health, Education and Welfare 1972; S. Tilkin, Chicago Kidney Foundation, personal communication).

It seems obvious, then, that it is more cost-effective to prevent human and environmental damage by severely restricting or eliminating the toxicants introduced into our environment than it would be to deal with the diseases and disorders that will result if we do not do so.

Implementation of the Consent Decree

Our final comments deal briefly with EPA's development of WQC for the 65 pollutants and classes of pollutants which are part of the Consent Decree to which CBE is a party. Under the terms of the Consent Decree, Best Available Technology (BAT) is to be installed by industry to control the discharge of the designated 65 pollutants. But the Decree also stipulates that more stringent regulations must be promulgated by EPA when BAT alone will be insufficient to protect public health and the ecosystem. Decisions to promulgate such strict regulations will be made largely on the basis of the WQC prepared by EPA.

In November 1977 we received the first drafts of the criterion documents written by EPA. We reviewed these documents with the aid of other scientists from around the country. It became obvious to us that there were serious deficiencies in the reports—including, but not limited to, a failure to include much of the available physical/chemical, environmental, and toxicological data and a failure to adopt a rational and consistent policy for deriving WQC from the data base. Because of the scathing criticism received by EPA from within the agency itself and also from outside reviewers, the agency was compelled to revise these documents. As a result, the agency was unable to meet its June 30, 1978, deadline, imposed by the court, for promulgation of the WQC.

Although at this time we have not seen copies of the revised criterion documents, we believe that, because of the methodologies employed in the revision, the revised WQC will not protect man and the ecosystem at large, as required under the law.

Criteria for Protection of Public Health

When the revised criterion documents are published this year, public health standards will be proposed for only 20 of the 65 pollutants involved. Eighteen of these 20 toxins were chosen, because they are known or presumptive carcinogens. The 20 toxins were also chosen, in part, because of the availability of a "best knowledge base" consisting of scientific health effect review documents.

For the 20 toxins, EPA has stated that the agency will concentrate on reviewing the carcinogenic effects as justification for establishing WQC. Thus the agency will not achieve a comprehensive review of the known health effects of any one pollutant. This policy is obviously in violation of the Consent Decree and with EPA's mandate under Section 304 of the Federal Water Pollution Control Act. As stated in the Consent Decree, water quality criteria published by EPA must "accurately reflect the latest scientific knowledge on the kind and extent of *all* identifiable effects on human health" and must state the "recommended maximum permissible concentrations (including where appropriate zero), consistent with the protection . . . human health."

This policy is also scientifically invalid. First, EPA seems to assume that for the 45 pollutants for which public health standards will not be developed, water quality criteria proposed to protect aquatic health will also protect humans. This is in spite of the fact that human health risks were the most important considerations in the selection of the list. Dr. Ian Nisbet, a consultant to the Task Force that selected the 65 pollutants, has estimated that "human risk considerations, if properly applied, would lead to more stringent criteria and standards than aquatic considerations for between 48 and 57 of the 65 pollutants" (Nisbet 1978).

Also, it is imperative that more attention be given to pollutants that are not carcinogens—pollutants whose ultimate effects, such as birth defects, mutations, nerve damage, and systemic, behavioral, and biochemical disorders, are as irreversible as cancer.

Criteria for the Protection and Propagation of a Balanced Ecosystem

Two methodologies are now being proposed for creating water quality criteria that would protect organisms other than humans in the environment. The first, developed by the Criteria and Standards Division within EPA, is similar to the methodology employed by EPA in the past in developing WQC. It attempts to predict "safe" levels by utilizing application factors with acute toxicity data and safety factors with chronic data and experimentally derived MATC's. The second, developed by the Office of Research and Development of the EPA, derives safety (sensitivity) factors by statistical methods. These sensitivity factors are to be applied to the LC50 or MATC of the species of median sensitivities in order to estimate a concentration that will protect 95% of the species.

While both of these methods possess certain advantages and disadvantages, both suffer from general features that are inconsistent with the regulatory position we have presented in this discussion.

For example, WQC must protect human health and aquatic animals *as well* as other members of the ecosystem. More attention must be given in both of these methodologies to the protection of other living organisms, such as plants, and to non-human consumers of aquatic life.

Also, although the sensitivity and safety factors employed in these methods may be appropriate to protect a single species in the laboratory, it is unlikely that they will adequately predict "safe" levels in the environment. They fail to consider sensitive and endangered species in the environment for which test data are unavailable, interactions among members of a species and within communities, and additive and synergistic effects with other pollutants.

Summary

In developing water quality criteria, a data base must first be compiled that includes all the relevant information on a pollutant, including physical/chemical properties, environmental fate and transport, and toxicological effects. The criteria formulated from such data must be conservative enough to account for the inherent unknowns in our understanding of toxicology and of the action of pollutants in the environment.

Unfortunately, the present methodologies employed by EPA in the development of WQC pursuant to the Consent Decree are inadequate to result in criteria that will be protective of man and the environment.

References

BURKS, S. L., AND J. L. WILHM. 1977. Bioassays with a natural assemblage of benthic macroinvertebrates. Pages 127–136 *in* F. L. Mayer and J. L. Hamelink, eds. Aquatic toxicology and hazard evaluation. ASTM STP 634. Am. Soc. Test. Mater., Philadelphia.

EPSTEIN, S. 1974. Environmental determinants of human cancer. Cancer Res. 34:2425–2435.

FISHER, N. S. 1974. Effects of PCBs on interspecific competition in natural and gnotobiotic phytoplankton communities in continuous and batch cultures. Microb. Ecol. 1:39–50.

HANSEN, D. J. 1976. Techniques to assess the effects of toxic organics on marine organisms. Pages 63–76 *in* U.S. Environ. Prot. Agency EPA-600/3-76-079.

MARKING, L. L. 1977. Method for assessing additive toxicity of chemical mixtures. Pages 99–109 *in* F. L. Mayer and J. L. Hamelink, eds. Aquatic toxicology and hazard evaluation. ASTM STP 634. Am. Soc. Test. Mater., Philadelphia.

NATIONAL ACADEMY OF SCIENCES. 1973. Water quality criteria (1972). Pages 122–123 *in* U.S. Environ. Prot. Agency EPA-RE-73-033.

NEEDLEMAN, H., AND S. PIOMELLI. 1978. The effects of low level lead exposure. Nat. Resour. Def. Counc., New York.

NISBET, IAN. 1978. Letter to Richard Stanford, U.S. Environ. Prot. Agency (dated April 7, 1978).

RICE, D., AND T. HODGSON. 1978. Social and economic implications of cancer in the United States. Paper presented to the Expert Comm. Cancer Stat. at the WHO/IARC meeting, Madrid, Spain (June 20–26, 1978) by the Natl. Cent. Health Stat.

U.S. DEPARTMENT OF HEALTH, EDUCATION AND WELFARE, PUBLIC HEALTH SERVICE. 1972. Statistics and epidemiology of lead posioning. FY:72–11.

Data Needs in Developing Water Quality Criteria

Leonard J. Guarraia, John J. Carroll, and Kenneth M. MacKenthun

Criteria and Standards Division
Office of Water Planning and Standards
US Environmental Protection Agency
Washington D.C. 20460

Abstract

Historically, the ambient water quality criteria have evolved from the initial publication in the early 1900's on effects of gas wastes on fish to the complex documents presently being developed on toxic pollutants defining ambient water quality criteria. The necessity for defining aquatic toxicity has placed a severe burden on the extand data base. To date, little coordinated effort to develop data on freshwater and marine organisms at comparable phylogentic levels for the same pollutant has occurred. A further understanding of the physical and chemical interactions of a pollutant and the aquatic system is required to define hazard to aquatic organisms and nonaquatic organisms. Human health effects resulting from exposure through the food chain and drinking water and from intermedia transfer are critical.

As the criteria considerations and scope have changed, so have the regulatory implications of the criteria. Since water quality criteria form the basis for state water quality standards, the criteria impact on the National Pollution Discharge Elimination Systems permits, serve to define the quality of source waters for water supply, and are the basis for all other water regulatory activities by the Environmental Protection Agency.

In their efforts to determine acceptable concentrations of water pollutants for different water uses, scientists have taken various approaches through the years. In this country, an initial effort to define water quality took the form of "effect levels" for a variety of pollutants on aquatic life (Shelford 1917). Ellis (1937), working from the literature, collected data on the effects of 114 pollutants on aquatic life and published a rationale for using standard test animals. Subsequently, attempts were made to standardize bioassay procedures.

In 1952, the State of California published the document "Water Quality Criteria" which summarized criteria promulgated by the state and interstate agencies as well as legal application of such criteria. Major beneficial uses were defined for eight categories and water quality criteria for protection of each use were identified.

The 1952 edition was expanded by the second edition, edited by McKee and Wolf (1963). This edition brought together under one cover the world's literature on water quality criteria as of that date. These criteria were identified and categorized according to their effects on the several designated water uses (domestic water supply, industrial water supply, irrigation waters, fish and other aquatic life, shellfish culture, and swimming and other recreational uses). The tabulation presented a range of values.

As one would expect, anomalies in test results were seen. In several cases, concentrations reported by some investigators as not harmful to an aquatic organism were the same or greater than concentrations reported by others to be harmful to the same organism. Such apparent inconsistencies resulted from differences in investigative technique, characteristics of the water used as a diluent for the toxicant, the developmental stage of the test organism, variations in temperature under which the tests were run, and the age of the test organism.

In 1966, the National Technical Advisory Committee was commissioned by the Secretary of Interior to develop water quality criteria for specific waters. The "Green Book" report, published in 1968, documented water quality requirements for particular and defined water uses (Water Quality Criteria 1968). The report was intended to be used as a basic reference for standard setting by state water pollution control agencies. This represented a merging of experimental work and the judgment of recognized water quality experts to yield criteria that would protect the quality of the aquatic environment and designated water uses.

When the Federal Water Pollution Control Act Amendments of 1972 were in their formative stages, the Environmental Protection Agency contracted with the National Academy of Sciences and the National Academy of Engineering to review and revise the 1966 National Technical Advisory Committee's water quality criteria and to develop a document that would include current knowledge. The result of this endeavor was "Water Quality Criteria, 1972," a 592-page report of the Committee on Water Quality Criteria of the National Academy of Sciences and the National Academy of Engineers (Water Quality Criteria, 1972 1973).

Section 304(a) of the 1972 Act mandated the Environmental Protection Agency to develop and publish criteria for water quality accurately reflecting "the latest scientific knowledge on the kind and extent of all identifiable effects on health and welfare including, but not limited to, plankton, fish, shellfish, wildlife, plant life, shorelines, beaches, aesthetics, and recreation which may be expected from the presence of pollutants in any body of water including groundwater."

Such quality criteria should protect aquatic organisms against acute lethal effects and the more subtle, long-term effects that may develop through bioaccumulation of a material or from other causes. They should protect man and other carnivorous animals which may consume edible portions of aquatic life that may have accumulated toxic materials, and they should prevent the impairment of water uses by maintaining water quality.

If eggs fail to develop, or if the sperm does not remain viable, the species would be eliminated from an ecosystem because of reproductive failure. Physiological stress may make a species less competitive with others and may result in a gradual population decline or an absence from an area. The elimination of a microcrustacean that serves as a vital food during the early period of a fish's life could ultimately eliminate a fish from an area.

The phenomenon of bioaccumulation of certain materials may result in chronic toxicity to the ultimate consumer in the food chain or to the organisms in which such accumulation occurs. Fish may absorb lethal toxicants from their fatty tissues during periods of physiological stress or when food is in short supply. Birds eating contaminated aquatic organisms may lay eggs with shells so thin they break before hatching.

Toxic substances may be hazardous to the health of man when he consumes aquatic organisms with tissue containing their residues. Thus, criteria for quality water, which provide a factor of safety against such hazards, are essential to maintain acceptable ambient water quality.

Publication of the "Quality Criteria for Water" in 1976 was the initial step in the Environmental Protection Agency's effort to define water quality. Two events have been a powerful driving force in the evaluation of water quality criteria. These two events were the Settlement Agreement entered into by the Agency and various environmental groups and the passage of the Clean Water Act of 1977. The Settlement Agreement requires the Agency to develop ambient water quality criteria for 65 pollutants and their families and the Clean Water Act classifies these pollutants as toxic pollutants.

Discussion

Criteria are usually developed by one of two routes—human health considerations or aquatic toxicity. Both areas suffer from insufficient data for many chemicals. This problem is aggravated by the fact that data needs are accelerating as new chemicals are considered, while the scope of data required broadens with increased sensitivity to the pollutants' potential ecological effects.

It has been clear from the outset to all who deal with criteria development that to base a criterion on effects on aquatic populations, it is desirable to use numerous data points (preferably from chronic studies) at each of several phylogenetic levels in both marine and freshwaters. For the vast majority of pollutants, only a fraction of such a data base exists. For instance, acute toxicity data (96-hour LC50's) are far more plentiful than chronic studies. Moreover, most studies have been carried out in freshwaters, rather than marine. And for many substances named in the Settlement Agreement and in the

Clean Water Act of 1977, few toxicity data of any kind were found in the literature.

Chronic and acute toxicity information are used in defining a pollutant's characteristics. For example, if a given concentration of toxicant is harmless to aquatic organisms exposed to it for a lifetime, what effects will follow very short-term exposure to twice that amount? Ten times?

Most tests address the concentration of toxicant in the water column. This is critically important, but it is also known that few toxicants remain in the water column indefinitely. They become bound to sediments; they escape to air; they are degraded; they disappear into the fatty portions of fishes. How is one to take sediment-bound toxicant into account in formulating a water quality criterion? How does the sedimentation process itself affect the potency of a toxicant over the long and short term?

Criteria must be based on ". . . the latest scientific knowledge . . ." according to law. For now the data base is too limited and the reliable models too unsophisticated to take into account many such parameters. The greater the data base developed, and the better the available models, the nearer the criterion can be placed to the highest level which will have no unacceptable effects. The time is coming when we will have verified models employing many parameters which cannot yet be systematically handled. Meanwhile, it behooves us to make the best possible use of available approaches and to expand and make uniform the data base we must use.

Which parameters should be concentrated on? The physical and chemical properties influence the fate of a toxicant. Water solubility tells something about how much of a toxicant will be found in the water column and gives an indication, when considered along with vapor pressure, of how quickly it will be lost to air. Octanol/water partition coefficient tends to correlate with bioconcentration. Since it is a relatively inexpensive test, it can be used to screen many chemicals, so that only those with high partition coefficients need to be tested for bioconcentration in aquatic organisms. The most critical testing from the standpoint of aquatic ecology, though, is direct toxicity testing, particularly long-term testing. A sample testing protocol for acute toxicity is outlined below; 96-hour LC50 or EC50 tests are used except as noted:

Acute Toxicity
   Algae
      Freshwater (*Selenastrum capricornutum*)
      Marine (*Skeletonema costatum*)
   Invertebrates
      Freshwater (*Daphnia magna*) (48-hour test)
      Marine (*Mysidopsis bahia*)
   Vertebrates
      Freshwater (*Lepomis macrochirus*)
      Marine (*Cyprinodon variegatus*)

While there may not be agreement with the choice of test species, and no assertion is made that these are the best test species, they do serve as examples of organisms distributed in US waters and they can be dealt with by our available laboratories.

Full chronic studies with the above organisms will provide data on which to base water quality criteria using current models. Not only is more chronic testing needed, but test procedures should be uniform to allow comparison of data generated by different laboratories. In the absence of full chronic data, embryo-larval and egg-fry studies (which expose organisms during sensitive life stages) are substitutes. These tests can be performed more rapidly and less expensively than full chronic studies.

The second approach to criteria development rests on human health hazard. At first glance this seems far removed from the study of effects on aquatic populations, but there is a critical connection. If a substance is found to be extremely toxic to man, or to cause birth defects, mutations, or cancer, it can threaten human health only to the extent that it reaches man. And one of the most convenient routes to man is through the aquatic organisms or drinking water. Mercury and PCB's are good examples. For these reasons, criteria based on human health may depend on the relationship between toxicant concentration in water and in the tissues of edible aquatic organisms or in drinking water.

Field data on aquatic and terrestrial organisms can be valuable, but cannot substitute for laboratory studies where variables can be controlled, where studies can be carried out in a given time period and the results closely determined, and where costs are less. Epidemiological studies are important. But like field studies, epidemiological research is difficult and expensive and may or may not yield usable data. Properly designed laboratory studies *will* produce usable data.

Non-Aquatic Tests

Use of non-aquatic acute and chronic bioassays is playing an increasingly significant role in establishing ambient water quality criteria. Acute bioassays using the mouse or rat, as well as long-term carcinogenic bioassays using the mouse or rat systems provide useful data, but the time and expense of these tests make such data difficult to obtain. Screening tests employing an indicator bacterial strain (*Salmonella typhimurium*) coupled with either rat liver or human liver microsomal activation in mutagenic bioassays will give presumptive information as to the potential for carcinogenicity in mammalian species. While the bacterial mutagenic bioassay is not necessarily accepted as proof of carcinogenicity, it will at least allow for prioritization of pollutants which should be screened for carcinogenicity in the in vivo mouse or rat system.

The immediate future of water quality criteria rests in large measure on generating comparable laboratory toxicity, bioconcentration, and physical/chemical data on a large number of pollutants. The long-term future of criteria development also depends on better models for assessing the influences of environmental variables such as sedimentation, adsorption, volatilization, and the like on the responses of aquatic populations to water pollutants, and on their transport from water to man.

Conclusion

The task of developing water quality criteria, then, is one of determining that concentration of particular water constituent which will adequately protect people and the aquatic ecosystem while providing a factor for safety because conditions in nature often are extreme and because the data base from which the criteria are developed is limited. The number of identifiable species associated with the water environment is great. The potential number of pollutants associated with the water environment is very large. The number of species that have been tested against potential pollutants is very small. The task of criteria development is challenging.

There is little disagreement that criteria must be established at a level that will provide a water quality in which all important aquatic life can live and produce progeny that are not constrained by the chemical quality of their environment. In addition, such criteria must protect humans and other animals that drink the water or consume aquatic life as food. Synergistic reactions of two or more pollutants in water must be part of the criteria development considerations despite the fact that relatively little information regarding this phenomenon is available for most pollutants. Recent concerns with carcinogens, mutagens, and teratogens and with pollutants that tend to accumulate within the tissues of an organism also are significant factors in criteria formulation.

The information base from which water quality criteria are developed includes acute toxicity, chronic toxicity, bioaccumulation or biomagnification of pollutants from water, carcinogenicity, mutagenicity, and teratogenicity. This information is obtained from laboratory experiments in which aquatic organisms are exposed to known concentrations of a pollutant for various time periods, from feeding experiments, from dose injection experiments, and from field investigative data.

Water quality criteria that will be published for the 65 substances in response to the Settlement Agreement will be used in several Agency activities associated with water pollution control. They will provide the justification for the development of effluent guidelines to regulate the discharge of such substances from the 21 specified industrial categories, which is another requirement of the Settlement Agreement. They will be available for states to consider in the next revision process of water quality standards, which specify criteria to maintain and protect designated uses of water. They will be used in developing the scientific basis for the designation of substances as hazardous discharges to water pursuant to Section 311 of the Federal Water Pollution Control Act. And they will provide recommendations and information to area-wide wastewater management programs, lake restoration programs, programs for discharge of dredged and fill material, the protection of drinking water supplies, and other related environmental programs associated with water.

If soothsaying is an allowed luxury, we believe that the future of water quality criteria development will be along two major lines: (a) such criteria will continue to be required for more and more identified pollutants as time passes; and (b) such criteria will become more and more an inseparable part of regulatory activity. The time will soon come, we believe, when national criteria will be a requirement of state water quality standards. Society, more and more, is developing an environmental ethic that recognizes that some pollutants must not be allowed to escape to the nation's waterways.

References

ELLIS, M. M. 1937. Detection and measurement of stream pollution. U.S. Bur. Fish. Bull. 48:365–437.

MCKEE, J. E., AND H. W. WOLF. 1963. Water quality criteria. State Water Quality Control Board, Sacramento, Calif. Publ. 3-A.

SHELFORD, V. E. 1917. An experimental study of the effects of gas wastes upon fishes, with especial reference to stream pollution. Ill. State Lab. Nat. Hist. Bull. 11:381–412.

WATER QUALITY CRITERIA. 1952. State Water Pollut. Control Board, Sacramento, Calif.

WATER QUALITY CRITERIA. 1968. A report of the National Technical Advisory Committee to the Secretary of the Interior. U.S. Gov. Print. Off. Washington, D.C.

WATER QUALITY CRITERIA, 1972. 1973. Natl. Acad. Sci., Natl. Acad. Eng., U.S. Gov. Print. Off., Washington, D.C.

Discussion Synopsis— Water Quality Criteria

H. E. JOHNSON, *Chairman*

T. DUKE, M. FINLEY, P. HULL, A. JONES, G. F. LEE, AND P. SCHNEIDER

Historically, water quality criteria have been derived from many data sources including field monitoring programs, toxicity tests, and environmental fate investigations. Regulatory agencies, scientific organizations, and individuals have defined and used water quality criteria in somewhat different terms, often resulting in confusion and varied interpretations. The workshop participants emphasized a critical need for distinct definitions of the terms *water quality criteria* and *water quality standards*.

The terminology used by the US Environmental Protection Agency (EPA) and described in "Quality Criteria for Water," July 1976, US Environmental Protection Agency states:

> The word "criterion," represents a constituent concentration or level associated with a degree of environmental effect upon which scienfitic judgment may be based. As it is currently associated with the water environment it has come to mean a designated concentration of a constituent that when not exceeded will protect an organism, an organism community or a prescribed water use or quality with an adequate degree of safety. A criterion in some cases may be a narrative statement instead of a constituent concentration. On the other hand a standard connotes a legal entity for a particular reach of a waterway or for an effluent. A water quality standard may use a water quality criterion as a basis for regulation or enforcement but the standard may differ from a criterion because of prevailing local natural conditions such as naturally occurring acids or because of the importance of a particular waterway, economical considerations or the degree of safety to a particular ecosystem that may be desired.

The National Academy of Sciences and National Academy of Engineering (NAS—NAE) (1973) have defined *criteria* as: "The scientific data evaluated to derive recommendations for characteristics of water for specific uses."

The NAS—NAE state that:

> As a first step in the development of standards it is essential to establish scientifically based recommendations for each assignable water use.

If adequate criteria for recommendations are available and the identification and monitoring procedures are sound, the fundamentals are available for establishment of effective standards. It is again at this step that political, social and economic factors enter into the decision making process to establish standards. . . . in some areas the scientific information is lacking, inadequate or possibly conflicting thus precluding the recommendation of specific numerical values.

The NAS—NAE state the recommendations ". . . are meant as guidelines only, to be used in conjunction with a thorough knowledge of local conditions."

The approaches and definitions of water quality criteria by the EPA and the NAS—NAE have subtle but important differences. The EPA terminology most often considers the criterion as a "designated concentration of a constituent," while the NAS—NAE considers criteria as "the scientific data evaluated to derive recommendations." In the latter case, a specific numerical "recommendation" is stated, "if adequate criteria are available."

The NAS—NAE emphasize that establishment of a recommendation implies access to "Practical methods for detecting and measuring the specified physical, chemical, biological and aesthetic characteristics," and that sound identification and monitoring procedures are necessary for establishment of effective standards.

The discussion panel considered the EPA terminology the most widely used today, *but* many panel members felt that the definition of *water quality criteria* developed by the NAS—NAE provides the more appropriate basis for developing water quality standards and for use in hazard assessment.

Information Needed for Criteria Development

This section of the workshop was concerned with assessing the data base necessary to establish scientifically valid water quality criteria. In particular the papers and ensuing discussions considered (1) the adequacy of information that is currently used to establish water quality criteria and (2) additional needs, including research, for improving upon the derivation and implementation of water quality criteria.

Many of our efforts to formulate water quality criteria are limited by the present state of the art in assessing the impact of chemicals on aquatic life and in predicting the distribution of chemicals in aquatic environments. Clearly, the data considered desirable for establishing water quality criteria must frequently be reevaluated and revised to be consistent with improved experimental technology and our ability to analyze and interpret laboratory data.

Three categories of information are desirable for formulating criteria. First, the physical-chemical properties of a compound must be considered since these will influence compound distribution within compartments of aquatic ecosystem. Many of the physical-chemical properties essential for this purpose have been described previously (Cairns et al. 1978) and are not discussed here. Second, the physical-chemical properties characteristic of specific environments must be considered since these influence the ultimate fate, distribution, form, and biological availability of a compound. Although site-specific information is not generally applicable for criterion formulation at the national level, the use of these data in criterion implementation is obligatory to achieve maximum practical utility of the criterion process. Hence it is essential that the criterion document include guidelines to recognize and use environmental fate information. Third, appropriate toxicological data are required as the basis for many of the criteria. Table 1 summarizes the toxicological information considered by the participants to be desirable for scientifically defensible criterion formulation.

The information specified in Table 1 suggests the kind of data considered to be desirable in most instances. There are circumstances within which the physical-chemical parameters or acute toxicity data, singly or in combination, will obviate the need for chronic toxicity testing. For certain types of testing (e.g., toxicity to algae) there is a need for work on interpretation of the environmental significance of laboratory test results. Further, these suggestions represent the state of the art and should not be considered final recommendations for required data.

The degree of conservatism that is used in developing water quality criteria was a concern that developed from two opposite points of view. One viewpoint considered a need for additional conservatism by using a factor of safety to compensate for unknowns that may be inherent in extrapolating laboratory toxicity data to actual aquatic ecosystems. This safety factor should take into account the degree of uncertainty in the data with respect to potential exposure to degradation productions and joint action of multiple chemical contaminants. The factor of safety should consider also the lack of suitable surrogate test organisms which reflect the vulnerability of endangered species. Safety factors should be generated in a manner that is reasonable and consistent with the recognition that wide ranges in tolerance often attributed to organisms are caused in part by sampling and experimental error.

A second group of participants considered that the degree of conservatism inherent in present toxicity tests precluded the need for additional safety factors. Experimental conditions used in the laboratory (e.g., continuous exposure to constant concentrations) probably overestimate field toxicity conditions and provide an

TABLE 1.—*Toxicity data considered to be desirable for water quality criterion formulation.*

	Acute	Partial chronic	or	Full chronic	Bioaccumulation[a]
Freshwater[b]	Fish Invertebrate Algae	Fish		Invertebrate	Fish
Saltwater[b]	Fish Invertebrate Algae	Fish		Invertebrate	Fish

[a] As required by partition coefficient and toxicity data.
[b] As required by projected environmental distribution.

intrinsic factor of safety. Further, there are moderating factors in the field (e.g., suspended solids, sedimentation, and variable rates of input) which tend to limit the biological availability of chemicals, and thus provide additional conservatism. Past research indicates that toxicity of chemicals is often additive, but synergistic action is rare and is not known to occur at low concentrations near the no-effect level; antagonism of toxic action may occur more frequently than synergism. Finally, past experience has not shown evidence that field toxicity conditions were underestimated where safety factors were not used in the criterion process.

Additional Criteria Needs

To improve on current methods for deriving and implementing water quality criteria, the participants identified the following needs:

(1) *Toxicity test procedures based on exposure of organisms to intermittent and variable chemical concentrations.*—Current water quality criteria are primarily derived from toxicity tests conducted for extended periods of time utilizing constant concentrations of test chemicals. In the field, however, concentrations of contaminants are frequently changing as a result of variable discharge rates, mixing processes, chemical reactivity and irregular periods of runoff following precipitation. There is a need to develop criteria applicable to situations of varying contaminant concentrations based on toxicity tests with fluctuating concentrations of test chemicals for different periods of time.

Criteria should reflect data from testing regimes that simulate typical durations of exposure concentrations existing in field conditions. Methodology is needed to simulate actual chemical use patterns, as well as more critical physical properties of the aquatic medium. This information is especially important for proper assessment of chemicals that are highly toxic but are rapidly biodegraded or otherwise removed from the aquatic ecosystem.

(2) *Test procedures to assess the impact of chemical forms associated with suspended or sedimented particles.*—Many chemical contaminants of potential concern exist in natural waters in a variety of forms, some of which are not readily available to aquatic species. Sorption reactions in natural waters are of major importance in governing the environmental fate and impact of chemicals in the natural environment. A substantial research effort is necessary to determine the biological significance of chemicals associated with sediments and suspended particulates in water. These studies should be directed toward improving the data base available for formulating water quality criteria.

(3) *Guidelines for implementation of water quality criteria into water quality standards.*— The generation of meaningful and scientifically defensible water quality criteria is becoming increasingly complex as more sophisticated analytical and biological methods are used in research laboratories. The criteria are often based on subtle biological effects that occur at very low concentrations of chemicals in water— sometimes at concentrations that are below the limits of reliable analytical methods. Hence the implementation of water quality criteria into water quality standards will require much more sophisticated assessment procedures. This has important implications for regulatory agencies as they translate the criteria into enforceable standards. State pollution control agencies will need assistance in the future to effectively develop and enforce water quality standards.

Guidance in the implementation of water quality criteria should be given in the form of a manual that contains the following information:

(a) A comprehensive review of the literature pertaining to the particular water quality criteria.
(b) Sufficiently detailed information on the environmental chemistry and fate of the contaminant to enable monitoring and assessing the environmental impact.
(c) Reliable analytical methods for measuring concentrations of the contaminant at levels somewhat below the criterion level.
(d) Guidance on the development of enforceable standards when analytical methods are not available to reliably measure contaminant concentrations.
(e) Guidance on the development of enforceable standards where contaminant concentrations fluctuate above and below the criterion level. This is especially important where criteria are based on tests with constant, chronic exposure.

(4) *A rational scheme for selecting the priority of chemicals for water quality criteria development.*—There is an urgent need to develop a rational and technically valid plan for ranking the order in which chemicals are considered for water quality criteria development. The large number of candidate chemicals coupled with limited resources for criteria development places a high premium on the priority given to a specific chemical. The scheme should follow a scientific evaluation procedure that considers such items as toxicity, volume, distribution, and potential hazard to human health and the aquatic environment. Failure to develop a well defined and scientifically valid scheme will result in inefficient and misdirected use of the scientific resources necessary for effective water quality criteria development.

State of the Art and Research Needs

Considerable toxicological data have been developed with specific "indicator" organisms that are representative of certain aquatic species. The lethal and sublethal effects of chemicals on these species are fairly well understood in relation to water quality criteria, but additional efforts are necessary to understand these data in terms of the aquatic species not normally tested in the laboratory. It is not practical to develop toxicity tests with a large variety of aquatic species but rather we must better use our existing test procedures to predict chemical effects on a variety of organisms.

Research is necessary to develop new and sensitive methods for measuring effects on biological functions in aquatic species, e.g., inhibition of collagen and chitin formation, and tests for mutagenicity, teratogenicity, and carcinogenicity. Frequently, we are capable of measuring such changes but are limited in our ability to interpret the results for species in the natural environment. For example, algal toxicity tests have been performed for some time but the practical application of the data in formulating water quality criteria is particularly difficult. The Effective Concentration (EC50), where algal growth is inhibited or chlorophyll is reduced by 50 percent in laboratory tests, cannot be readily interpreted in terms of potential impact in aquatic ecosystems.

Particular attention should be given to understanding how toxicity may affect populations in the natural aquatic environment. If laboratory tests indicate a chemical will reduce egg and larval survival, then what magnitude of effect is acceptable? Can a population lose 25% of the larval production, for instance, and still remain viable? Further, will this loss have an important impact on other segments of the ecosystem? Can we apply existing fishery management techniques, such as calculations of maximum sustainable yield, to solve these problems?

Questions were raised in the discussion regarding appropriate criteria to be used as indications of the impact of chemicals on ecosystems. General support was given to measuring changes in the functional role of biological communities but identification of the most sensitive measures and translation of the data to water quality criteria have not been adequately addressed. Useful quantitative data on community and ecosystem functions are scarce, but ecological theory is advancing with the use of mathematical models to guide data collection and research activities. Progress in the development of improved criteria will require a continuous integration of the toxicological and ecological information.

It would be useful to develop *indices of ecological health* to improve on methods for protecting fish and wildlife resources. Such indices could be developed by using experimental data coupled with environmental monitoring. Experimental systems yield valuable predictive data and monitoring provides for early warning of effects as well as validation of water quality criteria. As an example, eggshell thinning in birds and the relative abundance of the more sensitive species have provided useful indices of environmental impact from certain chemicals. This approach will continue to have greater importance for other environmental concerns as we improve our understanding of ecosystems.

Conclusions

(1) Uniform, precise definitions of terms used for water quality criteria and water quality standards are necessary. The terms used by the US Environmental Protection Agency are considered the most widely used today but the defi-

nition of water quality criteria developed by the National Academy of Sciences—National Academy of Engineering provides a more appropriate basis for developing water quality standards and for use in hazard assessment.

(2) The data base required for establishing scientifically defensible water quality criteria must include the physical-chemical properties of the chemical and toxicological data with suitable test species. Effective use of the criteria require data on the environmental chemistry and fate of the chemical and consideration of the physical-chemical properties of the environment.

(3) The toxicological data desirable for establishing water quality criteria includes acute toxicity with fish, invertebrates, and algae; partial chronic (egg-larval) toxicity tests with fish; chronic toxicity tests with invertebrates; and bioaccumulation tests with fish (if indicated by partition and toxicity).

(4) The degree of conservatism incorporated in the current water quality criteria was considered adequate by some participants and inadequate by others.

(5) The development and implementation of water quality criteria should be improved by including: (a) toxicity test procedures that are based on exposure of organisms to intermittent and variable dose concentrations that simulate field conditions; (b) improved test procedures to assess the biological availability and impact of chemicals associated with sediments and suspended particulate materials; and (c) guidelines for implementation of water quality criteria into enforceable water quality standards. The guidelines should include information on environmental fate, analytical methods, and procedures for monitoring.

(6) There is an urgent need to develop a rational, technically valid scheme for ranking the order in which chemicals are to be considered for water quality criteria development.

(7) The present state of the art for developing water quality criteria is limited by our ability to translate toxicity data from laboratory studies to population, community, or ecosystem effects. Research efforts are needed to improve the integration of toxicological and ecological information and their application to water quality criteria development.

References

CAIRNS, J., JR., K. L. DICKSON, AND A. W. MAKI, EDS. 1978. Estimating the hazard of chemical substances to aquatic life. ASTM STP 657. Am. Soc. Test. Mater., Philadelphia. 278 pp.

NATIONAL ACADEMY OF SCIENCES AND NATIONAL ACADEMY OF ENGINEERING. 1973. Water quality criteria—1972. U.S. Environ. Prot. Agency, EPA-R3-73-003, March 1973.

SUMMARY AND CONCLUSIONS

Summary and Conclusions

C. M. FETTEROLF, *Chairman*

W. BRUNGS, J. F. JADLOCKI, AND R. PARRISH

The Workshop Summary Committee has organized this encapsulation as follows:

Session I: *Background, objectives, and charge*
Session II: *Hazard assessment approaches*
  (A) Essence of discussion initiation papers
  (B) Observations and conclusions on hazard assessment approaches
Session III: *Hazard assessment philosophy and principles*
  (A) Essence of discussion initiation papers
  (B) Observations and conclusions on hazard assessment philosophy
Session IV: *Water quality criteria*
  (A) Essence of discussion initiation papers
  (B) Observations and conclusions of water quality criteria

The significant findings, observations and conclusions of these individual sessions are therein.

Session I: Background, Objectives, and Charge

Coordinators J. Cairns, Jr., (Virginia Polytechnic Institute and State University), K. L. Dickson (North Texas State University), and A. W. Maki (Procter & Gamble Co., Ohio) introduced the workshop by covering background, charging the participants, and identifying objectives.

They stated there is a continuing need for development of relevant testing methodology and integrated testing programs for effectively and efficiently assessing the potential hazards of chemical substances to aquatic life. In response to this need an ad hoc group planned and organized a workshop, "Estimating the Hazard of Chemical Substances to Aquatic Life," held June 13–17, 1977, at Pellston, Michigan (Cairns et al. 1978).

The participants of the Pellston workshop, following their review of several hazard assessment approaches, recognized that it would be beneficial to hold a subsequent workshop to examine additional approaches used in the United States, as well as those developed in other countries. They also suggested consideration of two related topics: (1) How well do laboratory-derived data on toxicity predict ecosystem effects? and (2) What constitutes an adequate data base for water quality criteria? The participants concluded that a subsequent workshop addressing these topics would be of significant assistance to professionals in industry, government, and environmental groups.

After Pellston, it was recognized very quickly that the workshop evaluation of the state of the art was having a profound influence on the hazard evaluation development process and its applications in industry and government. To implement the mandate from the participants, an ad hoc group commenced planning the 1978 workshop at the American Society for Testing and Materials (ASTM) meetings in Cleveland, October, 1977. The ad hoc group and the workshop coordinators identified potential participants, keeping in mind broadening the representation.

Coordinators' Charge

At Waterville Valley, the coordinators challenged the participants to represent their professions and not the particular organizational viewpoints associated with their jobs as they contributed to the workshop. Participants were encouraged to identify those aspects which they endorse as professionals and believe are usable in protecting aquatic life, and to subject the workshop topics to critical evaluation without losing sight of the workshop goal of producing a summary document with conclusions and recommendations for discussion by the participants on the final day.

Objectives

The participants were asked to:

(1) critically review and analyze the state of the art for assessing the hazards of chemical substances to aquatic life;
(2) critically review and analyze the state of the art in developing water quality criteria (recommendations) for safe concentrations; and
(3) evaluate the efficacy of laboratory-derived data to predict ecosystem effects.

Invited Charge

C. Fetterolf, Jr. (Great Lakes Fishery Commission, Michigan), in further charging the participants, emphasized that all persons involved with water quality management should recognize their multiple responsibilities, not only to their employers, but to the environment and its user groups, such as industry, local units of government, and the public. He stated that advocates of each of the competing users and the federal and state governmental agency representatives in attendance should not judge each other too harshly. He encouraged the participants to take off their "special interest hats," and figuratively slip on the "other Indian's moccasins" during the workshop, the better to understand each other's viewpoints and achieve resolution. These admonishments surfaced repeatedly in discussions.

To summarize the current environmental state of planet Earth, Fetterolf paraphrased Christman (1978) that modern man can press the world's absorptive limits, and in so doing has created a kind of molecular litter on a global scale, but that no hard evidence exists that humankind has suffered significantly from this action. The worrisome question is whether we will be able to recognize harmful global effects prior to the onset of irreversible change in some planetary function. Calling attention to the three objectives of the workshop, he charged the participants to recognize the review, analysis and evaluation hazard assessment as important steps toward development of predictive capability which can prevent those irreversible changes.

Session II: Hazard Assessment Approaches

Essence of Discussion Initiation Papers

J. Duthie (Procter & Gamble Co., Ohio) explained the authors had been charged to deal with the "review" portions of workshop objectives 1 and 2, concentrating on establishment and evaluation of the environmental risk associated with the introduction of a new chemical or continued use of an existing chemical, and the subsequent meshing of hazard assessment data with water quality criteria to determine what is safe and compatible with the ecosystem.

He charged the audience, and particularly the Session Synthesis Committee, to pick from the hazard assessment approaches to be presented, the common, positive pieces as they apply to how others will perform hazard evaluation for a variety of chemicals with different behavior.

C. Lee (Unilever Research, England) reviewed the diagrammatic representation of a sequential hazard assessment procedure, relating estimates of biological effect concentrations for a chemical to estimates of resulting environmental effect concentrations, which became a central theme and logo of the Pellston workshop. The synthesis illustrates that as a candidate material passes through sequential testing stages of increasing complexity to establish increasingly refined estimates of the biological no-effect concentration in relation to expected environmental concentration, the confidence limits narrow and become increasingly precise. Such firming is an objective of risk analysis, and demonstrates that as the confidence limits narrow, a smaller margin of safety between effect concentrations and expected environmental concentrations can be accepted.

He described a hazard evaluation scheme for the components of domestic detergent products which is based on predictive testing limited by time and resources. Using a potential detergent builder as an example, he explained how results of exposure, biodegradability, treatability, and toxicity tests determine the material's progress through the evaluation scheme.

Lee, who represented the most imaginatively named subgroup among the participants, the Bioconsequences Section of Unilever, posed a question which came up repeatedly throughout every phase of the workshop—where is the line drawn between clearance tests and confirmatory tests—when do you have enough information to stop—where does practicality end and superfluity start?

"Trigger" emerged as the buzz word in Session II and carried over into Session III as both authors and participants sought to identify the levels of effect or behavioral patterns of candidate materials which triggered the decision for continuation or termination of the hazard assessment process. Said another way—triggers identify criteria for decision and determine protocol for the decision sequence. The search is for unambiguous tests that will allow early decisions.

P. Lundahl (I.R.C.H.A., France) explained that strategy for hazard assessment of new chemicals in France is in the early development stage, with no firm protocols yet established. He stated that while a standard scheme could be extremely useful, a rigid pattern of tests would be inefficient because decisions to discard or commercialize a product can often be made without a large number of tests.

J. G. Pearson (US Army, Maryland) described hazard assessment for munitions-unique compounds and the evolution of sequential comprehensive strategies to determine long-term "safe" concentrations for both man and aquatic organisms. The title for his tabulation of triggers, "Criteria for not recommending a water quality criterion to EPA," caused the participants to acknowledge and discuss the longstanding differences in interpretation of the words "water quality criteria" and their relationship to standards, objectives, guidelines, recommendations, etc. The participants placed responsibility for clarifying the issue upon the US Environmental Protection Agency (EPA).

R. Lloyd (Ministry of Agriculture, Fisheries and Food, England) advised hazard assessors to make better use of easily accumulated data. For his example he chose the mean period of survival times throughout a series of acute tests plotted against the concentration to produce log-log concentration-response curves. Without such plots, the product of the tests is simply a 96-hour LC50. With such plots, the shape of the curve allows comparison with related test animals and chemicals as well as an early prediction of results for additional longer term concerns such as bioconcentration and chronic toxicity.

C. Walker (EPA and US Fish and Wildlife Service, Washington, D.C.) admonished that if test results are to be accepted and environmental consequences predicted with confidence, test organisms should be sensitive to the test substance, amenable to laboratory testing, geno- and phenotypically comparable to those in wild populations, ideally representative of species economically or ecologically important, and of known history. Timing constraints precluded submission of his paper for peer review and publication.

K. Fujiwara (University of Tsukuba, Japan) described the efforts of Japanese industry and government agencies to comply with a 1973 law on the regulation and production of chemical substances, with emphasis on testing for biodegradability and bioaccumulation. After four years of testing, 216 of 559 new chemical substances have been designated nonhazardous and 181 of 324 existing substances declared safe. He concluded the greatest challenge is to improve techniques for predicting the behavior of chemicals in ecosystems based upon laboratory simulations.

Several authors emphasized that hazard assessment plans should achieve approval to discharge, use, or market a material characterized as having low risk, but that plans designed to pass only those materials characterized as having no inherent risk would require too much testing. If this philosophy is accepted, the importance of risk evaluation in hazard assessment increases. There was general agreement that scheme design must retain a degree of flexibility and be tailored to the basic nature and projected use of the material.

Observations and Conclusions, Hazard Assessment Philosophy

The Synthesis Committee added two schemes, those of the American Society for

Testing and Material (ASTM) and the Federal Insecticide, Fungicide and Rodenticide Act (FIFRA) to those under examination. Approximately sixteen parameters were grouped into three categories of scope, design, and utility to facilitate the process of comparing and contrasting the schemes. In this manner, the differences and similarities of seven hazard schemes were reviewed to examine the development of hazard assessment schemes at the international level, and to influence future developments by highlighting certain evolving areas. This exercise, the papers, and discussion led to the following conclusions.

(1) Hazard assessment schemes exist throughout the world which provide a basis for the logical testing of the environmental fate of chemicals and their potential toxic effects on aquatic species.

(2) While hazard assessment schemes from various nations vary in their particulars, there clearly exists a similarity in several design features.

(3) A universal feature of these schemes is the concept of comparing expected or known environmental concentrations with biological effect concentrations of chemicals for reaching decisions about safety factors and the need for further testing.

(4) Virtually all schemes are tiered or phased in structure such that provision is made for sequential review and decisions concerning further testing.

(5) In spite of the similarities in the overall schemes, differences in test prescriptions could become burdensome in the international chemical trade, i.e., differences in test species or other specific test requirements could result in expensive duplication of effort and costly delays which may otherwise be avoided.

(6) Single-species acute and full life-cycle testing are generally the toxicological tests of choice among the various schemes.

(7) The degree of necessary sophistication for evaluating the environmental significance of a given chemical was the subject of some debate. Discussions focused on whether existing tests or more sophisticated state-of-the-art methods offer the best and most practical route to predicting environmental significance. However, there was general agreement that research for improved test development should be an ongoing, integral part of the overall effort to better understand the environmental consequences of today's chemicals.

(8) There was a reluctance to use a single number, e.g., LC50 or degradation rate constant, as a decision criterion within the testing schemes. Rather, it was generally agreed that the professional judgement of those using the schemes was necessary to decide if sufficient data were available or whether further testing was necessary.

(9) Some parameters used in comparing the various schemes caused considerable overlap with the subject matter in Session III—Hazard Assessment Philosophy. In particular, the emphasis of individual schemes was often different because the philosophy of the user was predicated on such subjective factors as regional or cultural influences. For example, the Japanese scheme was designed to emphasize the risks associated with bioconcentration of hazardous materials in fish tissue. In addition, the philosophy of what level of risk was acceptable and how this risk was quantified was discussed and evaluated for each of the schemes. For example, a lower level of risk often has the direct economic cost of conducting additional testing, and the decision to market a particular material with some known risk level often has to be weighed against the potential for that material to satisfy a market need, create jobs, and add to the overall standard of living. These subjects were subsequently considered in Session III during the discussion of "risk management."

(10) Some comments were received from the workshop participants after the Session II summary was presented on the final day of the workshop. These comments suggested that more attention might have been given consideration of specific tests which would facilitate making the best decision as early as possible in any given testing scheme. Two tests were suggested by participants as being the most meaningful and providing the highest probability of an accurate prediction of what a chemical's effect or no effect level will be. These tests, the embryo-larval test and the chronic life-cycle test, were recommended for use, but it was also noted that the choice of the specific test will *always* depend upon the characteristics and use patterns of the specific chemicals to be evaluated.

Session III: Hazard Assessment Philosophy and Principles

Essence of Discussion Initiation Papers

R. Kimerle (Monsanto Industrial Chemical Co., Missouri) updated his recent schematic diagram of a tiered sequential aquatic hazard evaluation program (Kimerle et al. 1978) to feature criteria to determine whether more data are needed, enough data are available, or the project should be terminated. He related the steps in the schematic to a narrowing of confidence limits with progression through tiers of testing—acute toxicity, additional acutes, short-term chronics, long-term chronics, field studies under experimental conditions, and, finally, field studies under use conditions. Said another way—screening, predictive, confirmative, and monitoring studies.

Kimerle stressed two basic principles of hazard evaluation schemes: knowledge of environmental fate plus toxicity; and progression of testing from individuals, to microcosms, to communities, to ecosystems. He expressed confidence in laboratory testing procedures, but questioned the design and results of field tests and the amount of confidence researchers have in them.

B. Neely (Dow Chemical Co., Michigan) stated that the task of experimentally making a hazard assessment for each potential new product, component thereof, and existing products and compounds is impossible without a preliminary screening program to predict potential environmental hazard. As Lloyd had done, he demonstrated the value of readily available "throw-away" data. By using computer modeling of such properties as molecular weight, vapor pressure, and water solubility, and comparison with benchmark performance of similar chemicals with known structure-activity correlations, the distribution of the chemical into the soil, air, and water compartments of the environment can be estimated for very little cost. In a computerized scenario involving chemical addition to the water compartment, the depuration rate from fish biomass can be estimated. He emphasized that if such a simple profile had been done for Kepone®, scientists would have been aware of its great potential to create environmental problems.

Neely's decision tree included a trigger involving confined use—if the product is to be used in a manner which prevents entry into the environment, obviously no further testing is required. The audience expressed strongly the need to define confined use and establish benchmark histories of products used in similar ways before agreeing that a product will truly be confined and not enter the environment.

E. Kenaga (Dow Chemical Co., Michigan) re-emphasized the theme previously expressed by Lloyd and Neely—make the best possible use of existing, readily available data—and demonstrated how statistically determined estimates of chronic no-effect concentrations for many chemicals can be obtained efficiently without extensive testing. Using known acute LC50's, MATC's, and AF's,[1] he developed predictive equations for combinations of related and unrelated organisms and chemicals. Early prediction of chronic toxicity can determine whether or not extensive testing appears necessary. He also demonstrated how to derive complementary theoretical data from a single known characteristic such as water solubility or soil adsorption.

J. Akerman (EPA, Washington, D.C.) together with D. Coppage (EPA, Washington, D.C.) explained the role of EPA in aquatic hazard assessment under federal laws (FWPCA, TSCA, FIFRA),[2] reviewed the agency's criteria for determination of unreasonable adverse effects, and iterated the data bases required for decision making on registration of general or restricted-use chemicals. Among the participants, greatest interest focused on the rebuttal procedures and provisions for economic-benefit decision criteria.

A. Maki (Procter & Gamble Co., Ohio) provided a great service to the participants by summarizing and reviewing the basic philosophical background, methodology, and utility of several published hazard evaluation schemes. He identified a troublesome problem in that similar materials, when run through a variety of schemes,

---

[1] Abbreviations: lethal concentration to 50% of test organisms, maximum acceptable toxicant concentration, application factor, respectively.

[2] Abbreviations: Federal Water Pollution Control Act; Toxic Substances Control Act; Federal Insecticide, Fungicide and Rodenticide Act, respectively.

provided a wide range of conclusions. He ranked thirteen schemes between the extremes "objective, clear decision criteria" and "subjective, no clear decision criteria," and generalized that those schemes following the most middle-of-the-road approach, providing well-defined triggers in early phases and offering more flexibility for scientific judgement in later phases, were the most effective.

Maki found there is wide disagreement among existing hazard evaluation schemes with respect to decision criteria triggering needs for additional test data. He concluded that until a larger comparative data base is available for many more real-world-tested chemicals, specific resolution and quantification of the most ecologically relevant or appropriate decision criteria are unlikely, especially in the later tiers of sequential assessment involving more sophisticated testing.

J. Hamelink (Eli Lilly Co., Indiana) proposed an industrial wastewater discharge scheme involving three major variables: river flow regime (log-normal distribution); chemical quantity discharge; and environmental fate kinetics. He stated that because all operate within definable limits of space, mass, and time, waste discharge need not be constant, but can be tailored to carefully predicted and monitored river flow. Laboratory studies can determine what concentration of a chemical over periods of time will or will not result in some measurable toxicity or tissue residue. In order to allow for tailored waste releases, he called for determination of pollutant concentration/response time relationships for sensitive components of the ecosystem combined with flow predictions for calculation of the probability of occurrence of the mean allowable pollutant concentrations. His recommendations, in part, call for site-specific water quality standards and flexible discharge volumes.

Observations and Conclusions, Hazard Assessment Philosophy

During Session III and the Session Synthesis Committee's activities, four conclusions from the Pellston workshop were advanced.

(1) Knowledge of the environmental fate of chemicals must be related to the concentrations of the chemicals known to cause adverse biological effects in order to perform hazard assessments. When the estimated environmental concentration (EEC) of a chemical is greater than the effect concentration (EC), further development of the chemical is questionable. When the EEC is less than the EC, it is reasonable to proceed with development of the chemical. When the EEC is the same as the EC, then more tests are required before a decision can be made.

(2) Tiered testing and data evaluation (sequential assessment) is a valid approach. The sequence of events is to screen, predict, confirm, and monitor.

(3) The fate of chemicals in aquatic environments can be predicted and is necessary for hazard assessment.

(4) Toxicity-testing methods are available for determining the effects of chemicals on aquatic life, but the confidence in the methods varies considerably.

To this base the Waterville Valley workshop participants added several observations and conclusions related to hazard assessment philosophy.

(1) The "benchmark concept" is useful in comparing new chemicals to those chemicals for which toxicity data exist and whose physical and chemical characteristics are known. The concept provides a basis for ranking new chemicals for toxicity and for predicting their fate and environmental concentrations.

(2) The estimated environmental concentration of a chemical is not fixed but varies according to input rates and environmental conditions. The concept of "risk management" was considered a valid approach toward mitigating environmental damages due to chemical usage. Risk management was considered a situation wherein the environmental concentration is controlled by various means (restricted use, waste treatment, controlled release, etc.) so that the environmental concentration is less than the known-effect concentration.

(3) Physical and chemical data provide a means of fate assessment which can be implemented early in the decision making process. Thus, requirements for such data should be in the first tier of a hazard assessment scheme.

(4) There is little basis for the prediction of the biological effects of chemicals based on simi-

larity of chemical structure. If the mode of action is known, however, this may be a basis of predicting biological effects.

(5) Based on existing knowledge, the results of laboratory tests on fate and effect can be related to fate and effect in the field. However, more resources should be devoted to (a) modifying existing laboratory methods so that they more closely reflect field conditions; (b) devising and implementing methods for testing aquatic birds and mammals when potential for exposure of these species exists; and (c) relating all known fate and effect data to chemical impact on aquatic ecosystems, not just on components of such ecosystems.

(6) Bioconcentration factors, rates of bioaccumulation and depuration, provide useful information to estimate damage to consumers of aquatic biota. Recent advances in estimating the bioaccumulative potential of chemicals through partition coefficient derivation or calculation and water solubility determination have provided additional important information for hazard assessment.

Session IV: Water Quality Criteria

Essence of Discussion Initiation Papers

In a paper linking Sessions III and IV, D. Mount (EPA, Minnesota) addressed the adequacy of laboratory-derived data to protect aquatic communities and discussed three common concerns: lack of testing on sensitive species; differences in stress factors between laboratory test animals and those in the wild; and the presence of other toxic chemicals in real-world water. He concluded that predictions based on laboratory data are as likely to overestimate as underestimate effects on aquatic communities in the field.

Mount, in his central theme, provocatively identified several differences in the perception by the public on one hand, and ecologists on the other, of success or failure of water pollution control and aquatic community management efforts. He stated ecologists visualize certain community structures and functions thriving in good water quality as a goal. He opined that the public did not perceive similar community structure and function as important, but that isolated features of communities, such as sport fish, are very important to the public. He emphasized that public attitudes should be recognized by ecologists because public support is crucial to the funding of both research and regulatory activities.

Mount challenged the participants with a series of curves portraying: (1) the decreasing probability of predicting effects on communities with laboratory data as the laboratory tests yield increasingly subtle indications; (2) the decreasing ability to separate contemporaneous community changes from those caused by stress as stress subsides; (3) the increasing tolerance (stability) of aquatic communities (good ones or bad ones, depending on viewpoint) as stress is reduced and the number of pathways for successful functioning increases; (4) the high value which the public places on specific community function versus low value on general community function, and the opposite values of ecologists; and (5) social value of small specialized communities and systems (e.g., a tailwater fishery), low social value for larger communities and systems (e.g., the river producing the tailwater fishery), and the very high social value ascribed to very large communities and systems (oceans).

L. Guarraia (EPA, Washington, D.C.) reviewed the history of water quality criteria development and EPA's current general philosophy toward application of such criteria in meeting the requirements of the Federal Water Pollution Control Act of 1972. He stated the need for expanded data bases for human health considerations, toxicity to aquatic organisms, and system interactions, stressing the impact water quality criteria have on all water quality regulatory activities of state and federal agencies.

Guarraia added that EPA is preparing a guidance manual for use by the states in transforming water quality criteria into standards as an attempt to achieve more uniform application.

K. Macek and S. Petrocelli (EG&G Bionomics, Massachusetts) titillated the participants and stimulated imaginations by examining, from the viewpoint of the fish, several aspects of criteria development and their "anthropocentric, piscicentric and ecocentric" objectivity. Discussants, characterizing themselves as ecosystem components (e.g., *Salmo salar,* rotgut minnows, starfish), emphasized selected perspectives, including neighborhood pride and the necessity for abundant and well-furnished bedrooms and nurseries.

The authors concluded that the objectives and derivation mechanisms were generally appropriate, but it was suggested that the risk levels, the degrees of conservatism utilized, and the ecological relevance of the criteria derived may be inappropriate.

D. Davoli (Citizens for a Better Environment, Illinois) argued for better data bases and more conservatism in application to take into account the inherent unknowns in toxicology, difficulties in extrapolating to the field, and the potential for enhanced toxicity due to environmental variation and multiple stress.

W. Brungs (EPA, Minnesota) explained both the philosophical basis behind the technical guidelines for deriving water quality criteria (US Federal Register, Vol. 43, No. 97, May 18, 1978) and the guidelines themselves. They describe a stepwise, rigid means of processing toxicological data to derive a water quality criterion.

Observations and Conclusions, Water Quality Criteria

The general agreement apparent during earlier sessions on hazard assessment and philosophy was lacking during the Water Quality Criteria session. This became readily apparent at the beginning when there was some difference of opinion even on the definition of the phrase "water quality criteria." In "Water Quality Criteria 1972" (National Academy of Sciences and National Academy of Engineering, 1973), the meaning is stated as "scientific data evaluated to derive recommendations for characteristics of water for specific uses." That is, criteria are the data base leading to a recommendation, not the recommendation itself. In "Quality Criteria for Water" (US Environmental Protection Agency 1977), EPA states that "Water quality criteria specify concentrations of water constituents which, if not exceeded, are expected to result in an aquatic ecosystem suitable for the higher uses of water." This latter definition identifies criteria as concentrations of toxic substances or other stresses and as a recommendation derived from a data base on effects.

Discussions relating to the basic data needs for the development of a realistic and defensible aquatic life criteria ranged from recommending the use of a very limited but well understood set of toxicity tests to recommendations that complex, ecosystem testing be required for criteria development. The former would result in the expansion of available knowledge on the use, interpretation, and extrapolation of existing methods such as embryo-larval procedures with fish. Proponents of the latter approach stated that ecosystem effects have been inadequately addressed in the development of virtually all criteria, and stressed that some multiple-species procedures are adequately developed for extensive use at present.

A discussion on the relative merits of a single, any-time-any-place criterion versus a dual criterion with mean and excursion-limited concentrations within a time frame found more support for the latter approach. Advocates stated it is a more toxicologically-defensible position because it is known that short excursions generally are acceptable in determination of chronic toxicity.

However, the development of a two-concentration, time-related criterion would require a greater data base. A need was defined for appropriate procedures (both mechanical and interpretive) for toxicity tests using intermittant exposure patterns, and for tests incorporating fluctuations in naturally variable conditions such as hardness, pH, temperature, and dissolved oxygen during continuous exposure.

As discussions progressed, the need for procedures to convert criteria (recommended conditions) to standards (legally required conditions) became increasingly obvious. Many were concerned that criteria would always be converted automatically into standards even though in "Quality Criteria for Water" (US Environmental Protection Agency 1977), Eckhardt Beck stated that criteria should be used with "considered judgement" based on the natural quality of water under consideration, the kinds of organisms that it contains, the association of those species to the tested species, and the local hydrologic conditions. Several attendees suggested a manual that would discuss, in depth, these and other considerations that should form the basis for that "considered judgement" to be used in the evolution of criteria into standards. The importance of physical, chemical, and biological fate in this process has been inadequately recognized and should become an important consideration.

The papers, discussion, and synthesis led to several conclusions.

(1) Because responsibility for administration of federal water quality laws rests with EPA, that agency must develop unequivocal definitions of water quality criteria and standards. At this workshop, the following definition was supported: water quality criteria are descriptions of chemical, physical, and biological characteristics of water that should protect defined uses. The criteria are developed from a scientific data base.

(2) Standards are not and should not be the same as criteria without adequate judgement that considers a variety of factors such as fate and local conditions.

(3) Criteria, and methods to derive them, should be dynamic and evolve as new relevant data become available.

(4) Available data bases should be very critically reviewed and synthesized to develop criteria. In the future, new types of data on multiple-species and ecosystem function tests may be necessary.

(5) Variability in toxicity related to such factors as intermittent exposure and variability in water quality should be studied. Information developed could provide additional data to make possible criteria incorporating mean and excursion-limited concentrations in a time frame, a system deemed much more defensible than single-number criteria.

(6) Effects data on aquatic birds and mammals should become a more important factor in criteria development.

(7) Analytical methods need to be improved so that field and effluent monitoring will become capable of measuring all water quality criteria concentrations.

(8) More care should be involved in selecting the chemicals for which aquatic-life criteria will be required.

(9) The US Environmental Protection Agency should develop a manual that addresses the development of water quality standards from water quality criteria.

References

CAIRNS, J., JR., K. L. DICKSON, AND A. W. MAKI, EDS. 1978. Estimating the hazard of chemical substances to aquatic life. ASTM STP 657. Am. Soc. Test. Mater., Philadelphia. 278 pp.

CHRISTMAN, R. F. 1978. No-growth revisted. Environ. Sci. Technol. 12(8): 867.

KIMERLE, R. A., W. E. GLEDHILL, AND G. J. LEVINSKAS. 1978. Environmental safety assessment of new materials. Pages 132–146 in J. Cairns, Jr., K. L. Dickson, and A. W. Maki, eds. Estimating the hazard of chemical substances to aquatic life. ASTM STP 657. Am. Soc. Test. Mater., Philadelphia.

NATIONAL ACADEMY OF SCIENCES AND NATIONAL ACADEMY OF ENGINEERING. 1973. Water quality criteria 1972. U.S. Environ. Prot. Agency Ecol. Res. Ser. EPA-R3-73-033.

US ENVIRONMENTAL PROTECTION AGENCY. 1977. Quality criteria for water. Off. Water Hazard. Mater., U.S. Environ. Prot. Agency, Washington, D.C. 256 pp.

Index

A

Acceptable daily intake values, 39
Act on the Control of Chemicals in France, 23
Act on the Protection of Nature, 23
Activated sludge, 52
Activated sludge test, 16
Acute flow-through toxicity tests, 42
Acute toxicity tests, 11, 39, 56, 58, 59, 102, 124, 140
    organisms for testing, 140
Algal toxicity test, 41
American Institute of Biological Sciences, 72
    hazard assessment scheme, 85
American Society for Testing and Materials,
    hazard assessment scheme, 89
Anaerobic biodegradability test, 20
Aniline, 53
Anthropocentric objectives of water quality criteria, 123, 124
Application factor equation, 106, 107
Application factors, 46, 59, 60, 101, 104–106, 121
    comparison of for fish, daphnids, and mammals, 109
Aquatic community function, 114, 115
    social and ecological value of, 115
Aquatic community structure, 112, 113
Aquatic community tolerance, 114
Aquatic ecosystem, 24
Aquatic ecosystem testing, 71
Aquatic field studies, 35
Aquatic hazard assessment strategy, 32
Aquatic test organisms, 102
Aquatic toxicity protocol, 39
Aquatic toxicity tests, 22
Aquatic toxicology, 83
Assimilative capacity, 1
Association Francaise de Normalization, 26

B

Basic data set, 4
Benchmark chemical approach, 75, 120, 153
Best practical treatment (BPI), 128
Bioaccumulation, 11, 26, 41, 50, 51, 61, 71, 139
Bioaccumulation test, 54
Bioconcentration, 22, 42, 43, 121, 124
Bioconcentration factor (BCF), 42, 43
Biodegradability, 9, 26
    potential tests, 9
    simulation tests, 9
    tests, 8, 13, 20, 24, 50, 52
Biomagnification, 26, 36
Brachydanio rerio (zebrafish), 26, 60

C

Calculation of environmental exposure concentration, 8
Cancer, 45
Carcinogenicity, 124
Carcinogenic risk assessment, 34
Cichlasoma nigrofasciatum (convict cichlid), 19
Citizens for a Better Environment, 132
Chemical fate, 24
Chemical sorption, 9, 10
Chemical stability, 9
Chemical transformation products, 42
Chlamydomonas variabilis, 26
Chlorpyrifos, 80
Chorella vulgaris, 22
Chronic toxicity, 22
    tests, 18, 39, 44, 56, 101–103, 124
Clean Water Act—1977, 139, 140
Comparative evaluation of hazard assessment schemes, 93–96
Compartmental analysis, 75
Compartmental modeling, 77
Consent Decree, 132, 139
Concentration-response curves, 58–60
Concentration-response relationships, 58, 59
Criteria definition, 143, 155
Critical life stage test, 124
Cyprinus carpio (carp), 50

D

Daphnia magna, 22, 26, 41, 44, 107
Daphnia sp., 13, 18, 22, 58, 101
Decision-making criteria, 120, 153
Decision tree, 14, 75, 76, 77
Decision triggers, 150
Detergents, 7
Dow Chemical Company hazard assessment scheme, 92, 93

E

Eastman Kodak Company hazard assessment scheme, 86, 87
Ecocentric objectives of water quality criteria, 125
Effluent chemical characterization, 34
Embryo-larval toxicity tests, 41, 43
Environmental chemistry, 34
Environmental effects, 32
Environmental fate reactions, 127
Environmental Protection Agency, 1, 31, 100, 127, 132
Estimated environmental concentration, 84, 119
Estuarine and marine die-away tests, 16
European Economic Community (EEC), 8, 66
European Inland Fisheries Advisory Commission (EIFAC), 10, 59
Expected environmental concentration, 74

INDEX

F

Federal Insecticide, Fungicide and Rodenticide Act, 68, 69
  hazard assessment scheme, 91, 92
Federal Water Pollution Control Act, 30, 68
  amendments of 1972, 139
Fish application factor, 44
Fish embryo-larval test, 70
Fish flesh tainting, 41, 42
Fish uptake/depuration rates, 121
Food and Drug Administration, 38, 45

G

Gammarus, 22
Gammarus pulex, 19
Ground water contamination, 25

H

Hazard assessment, 31
  approaches in France, 23
  case studies
    chlorpyrites, 80
    kepone, 78, 79
    mirex, 79, 80
    monochlorobenzene, 80, 81
  philosophy, 69
  principles, 69
  process, 26
  strategy, 30, 31
Hazard evaluation, 1
  principles, 7
  process, 84
H. J. Huek hazard assessment scheme, 90, 91
Human health effects, 32
Hydrological probability data, 128
Hydrolysis, 9

I

Indicators of chronic toxicity, 103
Invertebrate life cycle test, 71

J

Japanese Chemical Substances Control Law—1973, 50, 51
Japanese Industrial Standards (JIS), 52
Japanese Law on New Chemicals, 50
Japanese Ministry of Health and Welfare, 51
Japanese Ministry of International Trade and Industry (MITI), 51

K

Kepone, 78

L

Laboratory scale activated sludge plant test, 21
Lepomis macrochirus (bluegill), 41
LC50, 58, 101, 103, 121
LT50, 58

M

Macroinvertebrates, 36
Mammalian chronic toxicity tests, 39
Mammalian subchronic toxicity tests, 39
Mammalian toxicity studies, 38
Mathematical model, 33
Maximum Acceptable Toxicant Concentration (MATC), 43, 101–103, 121
Maximum Allowable Toxicant Concentration (MATC), 4
Maximum available toxicant concentration, 118
Methodology for effluent limits, 127–131
Microcystis aeruginosa, 41
Mirex, 79, 80
Monochlorobenzene, 80, 81
Monsanto hazard assessment scheme, 87
Munition compounds, 33
Mutagenicity, 124
Mutagenic screening tests, 38

N

National Academy of Sciences, 38
National Environmental Protection Act (NEPA), 23
National Pollutant Discharge Elimination System (NPDES), 128
National Research Council, 38
Nitroglycerin, 37
N-nitrosomorpholine, 34
No-effects concentration, 119
Nontarget organisms, 69

O

Octanol/water partition coefficient, 11
Organization for Economic Cooperation and Development (OECD), 8, 67

P

Paradox of the parabola, 131
Pellston Workshop, 2, 119
Periphyton, 36
Persistant chemicals, 51
Pesticide Registration Guidelines, 70
  additional tests, 70
  basic required tests, 70
  summary of physical chemical data requirements, 72, 73
Pesticides, 69
Pharmacokinetics study, 56
Photolysis, 9

INDEX

Piscicentric objectives of water quality criteria, 123–125
Piscitorial view of water quality criteria, 123
Plant growth test, 22
Poecilia reticulata (guppy), 26
Polychlorinated biphenyls (PCB), 50
Porous pot test, 21
Predicting community effects, 113
Predictive modeling, 75–77
Problem definition study, 33
Protection of aquatic communities, 112

Q

Quality Criteria for Water—1976, 139, 143
Quantitative decision criteria, 92, 93, 95–97

R

Relationship of value and aquatic community size, 115
Relevance of laboratory toxicity data for protecting aquatic communities, 116, 117
Replacement for phosphates in detergents, 11
Resource allocation method, 33
Richard Lloyd, hazard assessment scheme, 89, 90
Risk analysis, 1, 7, 8, 24
Risk assessment procedure, 45
Risk management, 119, 151
River die-away test, 16, 20
River hydrodynamics, 127

S

Safety factors, 45, 46
Scheme for prioritizing chemicals for testing, 147
Schemes for evaluating hazards, 85
Selenastrum capricornutum, 41
Semi-continuous activated sludge test (SCAS), 13, 20
Sequential hazard assessment scheme, 59, 74
  analysis of, 62–65
  characteristics of, 63
  comparisons of, 62
  decision criteria, 83–85
  international considerations, 66
  summary of, 62
  utility of, 62
Simulated ecosystem, 76, 77
Simulated field test, 71
Soil biodegradation test, 20
Solubility, 9
Specified chemical substance, 51
Standards definition, 143, 155
Stanford Research Institute International, 33
Static acute toxicity test, 39, 70
Stufenplan hazard assessment scheme, 88
Synergism of toxicants, 117

T

Teratogenicity, 124
Teratogenicity test, 56
Testing persistant chemical, 28
Tier testing, 74
TNT, 37
Toxic Substances Control Act (TSCA), 23, 68, 74, 83, 101, 102
  hazard assessment scheme, 92
  premanufacturing notification hazard assessment scheme, 93
Toxicity testing, 10
Toxicity tests, 24, 26
Toxicity tests with plants, 22
Treatability, 10
Trisodium carboxymethyloxysuccinate (CMOS), 7, 11
Tumorigencity test, 56

U

Unilever Company, 7, 8
Unilever hazard assessment scheme, 88
United States Army, 30

W

Water quality criteria, 2, 5, 30, 31, 33, 44, 46, 122, 123, 127, 143, 212
  analysis of criteria setting process, 125, 126
  criterion formulation, 135
  data base, 132
  derivation process, 123, 143, 144
  ecosystem protection, 136
  environmentalist viewpoint, 132
  nonaquatic tests, 141
  physical data, 133
  protection of public health, 136
  toxicity data requirements, 144
  toxicological and pharmacokinetic data, 133
  translation to water quality standards, 145
Water Quality Criteria—1952, 139
Water Quality Criteria—1968, 139
Water quality standards, 131
Waterville Valley Resort Association, 6
Waterville Valley Workshop, 119
  charge, 150
  conclusions, 148–156
  objectives, 149
World Health Association, 45
World Health Organization, 38